Recent Progress in
Japanese Speech Synthesis

Japanese Technology Reviews

Editor in Chief

Toshiaki Ikoma, *University of Tokyo / Texas Instruments Japan Ltd.*

Section Editors

Section B: Computers and Communications

Recent Progress in Japanese Speech Synthesis

by **Katsuhiko Shirai**
Waseda University, Tokyo, Japan

and **Masanobu Abe**
NTT Cyber Space Laboratories, Kanagawa, Japan

Gordon and Breach Science Publishers

*Australia • Canada • France • Germany • India • Japan • Luxembourg
Malaysia • The Netherlands • Russia • Singapore • Switzerland*

Amsteldijk 166
1st Floor
1079 LH Amsterdam
The Netherlands

British Library Cataloguing in Publication Data

Katshuhiko, Shirai
    Recent progress in Japanese speech synthesis. – (Japanese technology reviews. Section B : computers and communications ; v. 31)
    1. Speech synthesis – Japan
    I. Title II. Abe, Masanobu
    006.5′4′0952

ISBN 90-5699-095-0
ISSN: 1058-7306

Contents

Contents

Preface to the Series

Modern technology has a great impact on both industry and society. New technology is first created by pioneering work in science. Eventually, a major industry is born, and it grows to have an impact on society in general. International cooperation in science and technology is necessary and desirable as a matter of public policy. As development progresses, international cooperation changes to international competition and competition further accelerates technological progress.

Japan is in a very competitive position relative to other developed countries in many high-technology fields. In some fields, Japan is in a leading position: for example, manufacturing technology and micro-electronics, especially semiconductor LSIs and optoelectronic devices. Japanese industries lead in the application of new materials such as composites and fine ceramics, although many of these new materials were first developed in the United States and Europe. The United States, Europe and Japan are working intensively, both competitively and cooperatively, on the research and development of high-critical-temperature superconductors. Computers and communications are now a combined field that plays a key role in the present and future of human society. In the next century, biotechnology will grow, and it may become a major segment of industry. While Japan does not play a major role in all areas of biotechnology, in some areas such as fermentation (the traditional technology for making sake), Japanese research is of primary importance.

Today, tracking Japanese progress in high-technology areas is both a necessary and rewarding process. Japanese academic institutions are very active; consequently, their results are published in scientific and technical journals and are presented at numerous meetings where more than 20,000 technical papers are presented orally every year. However, due principally to the language barrier, the results of academic research in Japan are not well-known overseas. Many in the United States and in Europe are thus surprised by the sudden appearance of Japanese high-technology products. The products are admired and enjoyed, but some are astonished at how suddenly these products appear.

With the series Japanese Technology Reviews, we present state-of-the-art Japanese technology in five fields:

> Electronics
> Computers and Communications
> New Materials
> Manufacturing Engineering
> Biotechnology

Each tract deals with one topic within each of these five fields and reviews both the present status and future prospects of the technology, mainly as seen from the Japanese perspective. Each author is an outstanding scientist or engineer actively engaged in relevant research and development.

The editors are confident that this series will not only give a deep insight into Japanese technology but will also be useful for developing new technology of interest to our readers.

As editor in chief, I would like to sincerely thank the members of the editorial board and the authors for their contributions to this series.

TOSHIAKI IKOMA

Preface

How do we pronounce speech sound? This is the root of speech synthesis technology. To produce speech sound, human beings control their speech organs such as lips, tongue, jaw, and vocal cords in a complicated way. We can experience the complexity of speech organ control when we learn to pronounce foreign languages. The challenge to artificially generate speech sound has a long history. There is a record that in 1791 a mechanical tube was invented that could reproduce speech sound. Currently, however, digital signal processing is employed for speech synthesis. In the 1960s, speech synthesis technology made significant progress along with the development of a speech production theory; i.e., based on the theory, speech sound was synthesized using digital signal processing, and listening tests were performed to confirm the validity of the theory. Knowledge of the important characteristics of phoneme sound was accumulated, which resulted in a formant synthesizer. This is the advent of speech synthesis. The progress of speech synthesis technology is not independent from the astonishing progress in computer hardware and software. In the late 1980s, it was proposed to synthesize speech based on speech waveforms instead of speech spectra. In this approach, the entire speech signal of each phoneme is stored, and marks that indicate the starting time of every vocal cord vibration must be assigned for pitch-synchronous processing. For researchers whose research was limited by memory, the idea might seem inappropriate, but currently the approach is the most popular way of generating speech signals. According to the history of speech synthesis, it is true that the quality of synthesized speech increases with the amount of speech data available. From the 1970s, several practical applications of speech synthesis were introduced; the text-to-speech (TTS) systems that automatically convert text into speech sounds are especially interesting. We can see that speech synthesis technology is multidisciplinary and includes linguistics, phonology, phonetics, digital signal processing, mathematics, computer science and psychology.

The aim of this book is to introduce and explain the key technologies of speech synthesis. The main focus is the research performed in Japan over the period of 1987–1994. The basic trend of this period is the use of the statistical approach. The authors believe that this approach, well entrenched in Japan, will spread to other languages and will be the basis of future TTS systems. Speaking of the different disciplines, the progress in each area is not the same. The knowledge obtained through the speech production theory in the 1960s, for example, is still useful to develop speech synthesis systems. In this book, the basic knowledge offered by earlier systems is surveyed in every section.

Currently, state-of-the-art synthesized speech is understandable, but remains quite different from human speech. As future research, the expressiveness of synthesized speech including emotion might be important. The authors hope that the techniques explained in this book will help advance and stimulate research to new heights.

CHAPTER 1

Introduction

Speech is the primary communication tool for human beings, so to enable flexible and natural communication between humans and computers, speech synthesis is considered to be an essential technology along with speech recognition. Currently, there are two major approaches for speech synthesis; (1), concatenation of some human speech segments and (2) synthesis by rule. In the former, digitally recorded speech segments such as sentences, phrases, and words are stored and played back under computer control. This method is easy to achieve and has been widely used in the applications requiring only a limited number of expressions. However, in many applications, one cannot know in advance which words or which types of sentences may be needed. The technology of synthesis by rule makes it possible to generate any speech message, even those that human beings have never spoken before. Therefore, synthesis by rule offers more flexibility in terms of generating speech and is expected to become more popular. For the last 30 years, as a more specific application of synthesis by rule, text-to-speech (TTS) systems that automatically convert text messages into speech has been one of the most important goals in the field of speech synthesis. Thanks to recent progress in computer technology, TTS software commercial products are now available for personal computers. They are used in e-mail reading systems that enable a user to hear e-mail messages over the phone, voice navigation systems that speak the names of the town or the intersection where the car is currently located, text checking systems that help to find errors by reading the texts out, weather forecasting systems and so on.

Figure 1.1 shows a block diagram of a TTS system. Input texts are analyzed by referring to a word dictionary which contains phonetic transcriptions and linguistic information such as accent, part of speech and so on. Using the outputs of the text analysis, the prosody control block generates intonation patterns so as to best match the input texts by applying rules constructed in advance. Intonation patterns are usually

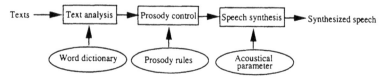

Figure 1.1. A block diagram of text-to-speech system.

parameterized by speech power, phoneme duration, and fundamental frequency. The speech synthesis block generates the speech signal; i.e., the speech signal is represented as a set of acoustical parameters based on a model and, by using digital signal processing, the speech signal is generated by referring to rules that map phonetic transcriptions to sets of acoustical parameters.

Text-to-speech conversion systems have been studied for many languages including English, Swedish, French, and German. Each system generates speech of only one language due to the unique characteristics of each language. Japanese strongly differs from European languages. The first aim of this book is to describe problems and solutions unique to Japanese TTS. In particular, the readers will find interesting problems in the linguistic analysis of Japanese in the text analysis block. Secondly, this book explains research performed in the period 1987–1994 in Japan with emphasis on TTS. Key items are statistical approaches and speech databases. The authors believe that these ideas, developed during the period, will be the key technologies for future systems and are applicable not only to Japanese, but also to other languages. To help readers understand the importance of the new ideas, each section starts by briefly reviewing conventional techniques. Thirdly, approaches to TTS evaluation are shown. Although evaluating TTS systems is important, little research has been conducted worldwide. Because many technical factors are involved at different levels and the subjective views of humans is hard to quantify, TTS evaluation remains difficult. The approaches described in this book will help the reader to understand these problems. Finally, research aimed at creating a new direction for speech synthesis is described. Because most users are not satisfied with the speech quality of current TTS software products, most research is aimed at improving speech quality. However, the last part of this book provides a fresh viewpoint and focuses on widening the diversity of speech styles available.

CHAPTER 2

Key Technologies for Text-to-Speech

2.1. Text Analysis

The aim of text analysis is to transform the input text into a sequence of phonetic symbols and to assign accent to utterance groups. All text analysis procedures deal with just symbols, not acoustical parameters. Because Japanese texts are written in a different manner from the European languages, some unique problems occur during text analysis. First of all, the Japanese writing system has both Chinese characters (Kanji) and Japanese syllabary (Kana), and most Kanji have several different readings. For example the character "行" has the following readings [Saito-92]:

行く /i ku/ < go >
行う /okona u/ < do >
行列 /gyou retsu/ < a procession >
銀行 /gin kou/ < a bank >
行燈 /aN doN/ < a paper-framed light >
　Even the same Kanji strings can have different readings according to context.

　For example:

初日 /shonichi/ < the opening day >
　　/hatsuhi/ < the sunrise on New Year's Day>
十分 /juppuN/ < ten minutes >
　　/juubuN/ < enough >
　Secondly, in Japanese texts, Kanji and Kana symbols can be mixed. There are several possible Kanji-Kana combinations for writing the same word. An example is

締め切り　締切り　締め切 /shimekiri/ < closing >
　Thirdly, there are no symbols that separate words or phrases, except for punctuation marks. The following text can be decomposed as follows.
日本語の音声が合成されます。/nihoNgono oNseiga gouseisaremasu/
< Japanese speech is synthesized. >
日本＄語＄の％音声＄が％合成＄され＄ます。

Here, $ indicates a morpheme boundary and % indicates an inter phrase boundary.

As we can see, Japanese texts are ambiguous at several levels. The first step of text analysis is morphological analysis; i.e., texts are decomposed into morphemes. In almost all text-to-speech systems, the process is performed sequentially from left to right using one or more dictionaries [Sagayama-83]. The dictionary contains morpheme strings, accent type, and grammatical information such as part of speech, semantic relations between adjacent morphemes and so on. Japanese parts of speech are noun, verb, adjective, adverb, auxiliary verb, auxiliary adjective, auxiliary noun, particle, prefix and suffix. The information in the dictionary should permit optimal morpheme decomposition. Input characters are compared with the morphemes in the dictionary from left to right, and a criterion is used to select the best morpheme. After finding the morpheme, the following characters are processed in the same way. To reduce the amount of computation, this process is performed by dynamic programming. Several criteria have been proposed. Examples are (1) to select the longest word, (2) to minimize cost for adjacent morphemes connection, (3) to minimize the number of morphemes in a sentence, (4) to give high possibility for important words which frequently appear in Japanese texts, (5) to give high possibility to Kanji, (6) to give high possibility to idioms.

Because Japanese has some exceptional Kana symbols and phonetic representation associations, after morpheme determination, the following rules are applied. [Sagisaka-82]

(1) Transcription of particles
     The particles "は" /ha/ < presupposition > and "へ" /he/ < goal > are transformed as follows:
     /ha/ → /wa/
     /he/ → /e/
(2) Transcription to long vowel
     The vowel "う" /u/ which succeeds a syllable including vowel /o/ is transformed to the long vowel symbol ":".
     "とうきょう" /toukyou/ → /to:kyo:/
     When two same vowels adjoin within a morpheme, they become a long vowel.
     "こおり" /koori/ → /ko:ri/

(3) Unvoiced vowel
    Vowels /i/ and /u/ are unvoiced, when they occur between voiceless consonants.
(4) Nazalization of /g/
    /g/ is basically nazalized and is represented by /ng/ except when /g/ is located in the initial position of a word.
(5) Syllabic nasal
    Syllabic nasal /N/ is pronounced in various ways depending on the following sounds. The rules are roughly summarized as follows.
    /N/ → /m/ : before p, b and m.
    /N/ → /n/ : before t, d, and n.
    /N/ → /ng/: before k, g, and ng.
    /N/ → /N/ : the final position and others.

Finally, accent nucleus position is assigned [Sagisaka-83] [Sagisaka-84]. The word accent in the dictionary is a sort of an intrinsic accent; i.e., the word accent applies only when the word is uttered in an isolated manner. In general, when a word is concatenated with other words or morphemes, its accent nucleus position changes. A set of rules for accent assignment has been proposed. The proposed rules determine accent assignment for compounds and accent phrases using the intrinsic accent and the accent concatenation style. Table 2.1 [Saito-92] shows these rules.

2.2. Prosody Control

It is well known that the acoustic features of speech have two aspects: segmental and prosodic. While the segmental features are quite local (such phenomena are held for about 5–20 ms) and mainly convey phonemic information, the prosodic features are rather global (such phenomena are held for about 1–3 s) and convey information such as accent, intonation, prominence, rhythm, tempo, and pauses that are strongly related to sentences. In speech synthesis-by-rule, the prosodic features are specified by fundamental frequency, phoneme duration and speech power. Because the phoneme number of any language is limited, almost all recent text-to-speech systems store phonemic information in the form of speech segment templates. However, the number of sentences is infinite, so it is difficult to store prosodic templates. For this

Table 2.1 Accent concatenation rules in intra-phrase accentuation. S. Saito (ed.) *Speech science and technology*, Tokyo Ohm Co. pp. 171 (1992).

Phase Configuration	Accent Concatenation Style	Accent Concatenation Example			
Stem-Particle or Stem-Auxiliary	(i) Subordinate	waru (laugh)	+	hodo	_ waruhoda (as far as laughing)
		arúku (walk)	+	hodo	arúkuhodo
	(ii) Dominant	waru	+	yo da▲da	_ waruyó▲da (seem to laugh)
		arúku	+	yo_da	_ arukumái
	(iii) Predominant	waru	+	mai	_ warumái (will not laugh)
		arúku	+	mai	_ arukumái
	(iv) Assimilating	waru	+	seru	_ warawaseru (make laugh)
		arúku	+	seru	_ arukaséru
Stem-Suffix	Standard	taisho▼ (symmetric	+	ten	_ taishó▲se (symmetric point)
	Deaccented	taisho▼	+	se▼	_ taisho▲se (symmetricity)
Stem-Stem	Generating	taisho▼	+	ido▼	_ taisho▲ído (symmetric transfer)
	Preserving	taisho▼	+	sayó▲so	_ taisho▲sayó▲so (symmetric operator)

reason, prosodic features are controlled by rules in speech synthesis-by-rule systems. In this section, we will discuss methodologies for controlling the fundamental frequency, phoneme duration and speech power.

2.2.1. Fundamental Frequency

Among the prosodic features, the fundamental frequency (F_0) contour is extremely important because of its many roles, such as indicating phrase boundaries in a stream of speech sound, to focus on words in

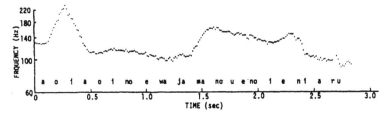

Figure 2.1. An example of a measured F_0 contour. S. Saito (ed.) *Speech science and technology*, Tokyo: Ohm Co. pp. 44 (1992).

sentences, speaker's intention and so on. Moreover, because Japanese is a tonal language, in Japanese speech synthesis-by-rule, F_0 contour also indicates the accent of words or phrases. Every Japanese word has a specific accent type that is specified by combinations of low-tone and high-tone. In some phoneme sequences, different accent type means different meaning, such as "ha (high-tone) shi (low-tone)" means "chop sticks", and "ha (low-tone) shi (high-tone)" means "bridge". Generating the appropriate F_0 contour is obviously very important in transmitting correct meaning as well as improving the understandability of synthesized speech. Therefore, intensive studies have addressed F_0 models that produce F_0 contour and F_0 generation rules that estimate F_0 contour using linguistic information.

Before explaining details of these F_0 models and F_0 generation rules, let us observe an example of F_0 contour. Figure 2.1 [Saito-92] shows the F_0 contour of the Japanese declarative sentence / aoiaoinoewajamanouenoieniaru/ ("The picture of a blue hollyhock is in a house on top of the hill"). The sentence was uttered with neutral declarative intonation. The F_0 contour generally has the following two characteristics. As clearly shown in Figure 2.1, the first characteristic is a local hump with rise-plateau-decay shape. This phenomenon is related to accent. The other characteristic which is not visibly drawn in the F_0 contour itself, is a baseline which gradually decreases toward the end of the sentence.

Any F_0 model for speech synthesis-by-rule tries to approximate these two characteristics. There are two approaches in terms of dealing with these characteristics: explicit and implicit separation of the local hump and the base line. Explicit separation models have a long history and

Table 2.2. F_0 models and F_0 generation rules.

Model	Hat pattern model	Fujisaki model	Syllable base model	Database template model
Separation of hump and baseline	Explicit	Explicit	Implicit	Implicit
Statistical algorithm	Tree-based approach	Multiple split regression	Quantification theory	

are employed in most of the recent synthesis-by-rule systems. One motivation behind the proposal of implicit separation models is the difficulty of separating the local hump from the base line in a one dimensional F_0 contour. Because of this difficulty, the separation is usually checked manually. This is a serious disadvantage when dealing with a lot of speech data to develop F_0 generation rules. In this section, we will discuss two explicit separation models: hat pattern model and functional model (Fujisaki model), and two implicit separation models: syllable based model and database template model.

In terms of developing F_0 generation rules, statistical approaches are becoming popular. The construction of large speech databases make this trend possible. Needless to say, the rapid progress in computer hardware and software is also a factor. Several statistical algorithms have been developed. In this section, we will explain three: a quantification theory, a tree-based approach, and a multiple split regression. Although the statistical algorithms can be applied to any kind of F_0 model, we will explain the algorithms using a particular combination of a F_0 model and a statistical algorithm as shown in Table 2.2. Before the statistical approaches were proposed, extensive studies were carried out to reveal the relationship between F_0 contours and linguistic information. Because the findings are important and are the basis of designing the factors employed in statistical approach, one section will explain the findings.

F_0 Model

Hat pattern model

In this model, the baseline phenomena and the hump phenomena are modeled as a basic intonation pattern and an accent component pattern,

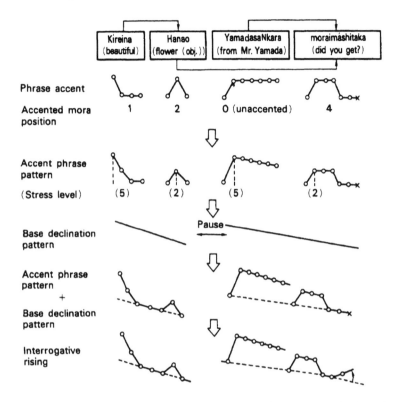

Figure 2.2. F_0 contour generation in sentence speech. S. Saito (ed.) *Speech science and technology*, Tokyo: Ohm Co. pp. 174 (1992).

respectively; the F_0 contour is generated by superposing both patterns [Hakoda-76]. Figure 2.2 shows a flow of F_0 contour generation. The basic intonation pattern is approximated by a straight line declining toward the end of a sentence. The accent component pattern is strongly related to the Japanese word accent which is usually represented by a tonal pattern of high-tone and low-tone that are assigned in every mora. Figure 2.3 shows an example of the word accent represented by tonal pattern [Kindaichi-81]. However, the accent component patterns can not be the same as the tonal pattern, because the F_0 contour represents physical phenomena. Therefore, experiments were carried out to clarify the relationship between the tonal patterns and the accent component

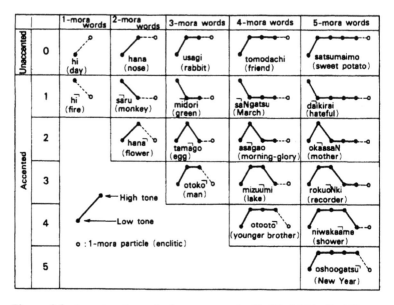

Figure 2.3. Accent patterns in Japanese words. H. Kindaichi, K. Akinaga, *Japanese accent dictionary*, Tokyo: Sanseido, (1981).

patterns. According to the results, the tonal pattern can be approximated by a sequence of F_0 values which are sampled at the center of gravity in each vowel segment. (called "point-pitch") [Hashimoto-74] and can be mapped to accent component patterns as shown in Figure 2.4 [Sato-77].

Consequently, the approximated F_0 contour is represented by simple linear interpolation of the point-pitch. This model is based on the idea that, from the perceptual point of view, point-pitch is the most important value and should be preserved. The hypothesis was supported by a perceptual experiment that was performed using synthesized speech [Hashimoto-74]. According to the experiment, in terms of quality there is little difference between speech synthesized from measured F_0 contours and speech synthesized from F_0 contours generated by linearly interpolating the point-pitch.

A functional model (Fujisaki model) [Fujisaki-71]

In this model, the baseline phenomena and the hump phenomena are

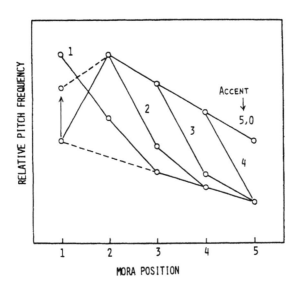

Figure 2.4. Point pitch patterns in Japanese 5-mora word. H. Sato, K, Hakoda, "Speech synthesis based on stored speech segments and rules (in Japanese)," *ECL Tech. Jour.*, **27**, 12. pp. 2551 (1977).

modeled as a phrase component and an accent component respectively, and the F_0 contour is generated by superposing both components [Hirose-79]. An interpretation of the model was reported on the basis of physiological and physical properties of the vocal cords and the structure of the human larynx. One unique idea in the model is that phrase and accent components are represented by different responses of a critically-damped second order system. Inputs to the system are called phrase commands and accent commands, and they are a set of impulses and a set of stepwise functions respectively.

Figure 2.5 shows the model that generates sentence F_0 contours. The F_0 contour is formulated as follows:

$$\ln F_0(t) = \ln F_{min} + \sum_{l=1}^{I} A_{P_i} G_{P_i}(t - T_{0_i})$$

$$+ \sum_{j=1}^{J} A_{a_j} \left\{ G_{a_j}(t - T_{1_j}) - G_{a_j}(t - T_{2_j}) \right\} \tag{1}$$

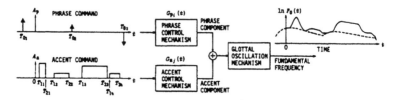

Figure 2.5. A function model for the process of generating F_0 contours. H. Fujisaki, H. Sudo, "Synthesis by rule of prosodic features of connected Japanese," *Proc. of 7th ICA*, Vol. **3**, pp. 133–136 (1971).

where

$$G_{P_i}(t) = \begin{cases} = \alpha_i^2 t \exp(-\alpha_i t), & \text{for } t \geq 0, \\ = 0, & \text{for } t < 0, \end{cases} \qquad (2)$$

and

$$G_{a_i}(t) = \begin{cases} = Min\left[1 - (1 + \beta_j t)\exp(-\beta_j t), \theta\right] & \text{for } t \geq 0, \\ = 0, & \text{for } t < 0, \end{cases} \qquad (3)$$

The symbols in Eqs. (1), (2), and (3) are

$G_{P_i}(t)$: the impulse response function of the phrase command,

$G_{aj}(t)$: the step response function of the accent command,

F_{min} : asymptotic value of fundamental frequency in the absence of accent components,

I : number of phrase commands,

J : number of accent commands,

A_{pi} : magnitude of the i-th phrase command,

A_{aj} : magnitude of the j-th accent command,

T_{0i} : timing of the i-th phrase command,

T_{1j} : onset of the j-th accent command,

T_{2j} : end of the j-th accent command,

α_i : natural angular frequency of the phrase control mechanism to the i-th phrase command,

β_j : natural angular frequency of the accent control mechanism to the j-th accent command,

θ : a parameter to indicate the ceiling level of the accent component (generally set equal to 0.9)

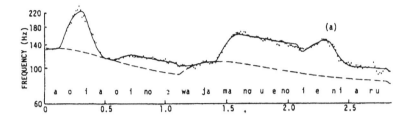

Figure 2.6. Analysis-by-synthesis of F_0 contours. S. Saito (ed.) *Speech science and technology*, Tokyo: Ohm Co. pp. 47 (1992).

All of the above parameters were estimated so that the model output matches measured F_0 contours. The problem is solved by the Analysis-by-Synthesis method, i.e., finding the model parameters that minimize the mean squared error between the measured F_0 contour and the model output on the F_0 logarithmic scale. Figure 2.6 shows an example of the F_0 contour approximation for the Japanese declarative sentence /aoiaoinoewajamanouenoieniaru/ ("The picture of a blue hollyhock is in a house on top of the hill"). The sentence was uttered using neutral declarative intonation. The dots in the figure indicate the measured F_0 contour, the solid curves indicate the best approximations generated by the model, and the curves plotted as dashed lines indicate the phrase components estimated at the same time.

Syllable based model [Abe-92]

The syllable based model was designed to avoid explicitly separating the baseline and hump phenomena. Because the baseline phenomena is not visible, a reasonable separation is difficult to achieve, especially in automatic separation. This difficulty is sometimes a drawback when using a large scale database for developing F_0 generation rules. Instead of explicit separation, the model implicitly represents the baseline phenomena by generating an F_0 value for each syllable position. The modeling seems reasonable because, in nature, the first syllable has

		SBU Attributes				Predicated Parameters		
		Consonant Class						FO
Syllable Pattern	Tone Pattern	Current Syllable	Following Syllable	Max. F_0	Syllable Position	Av. F_0 (Hz)	F_0 Diff. (Hz/3ms)	
/a/	–LH	no consonant	liquids	170 Hz <	1	100	0.4	
/ri/	LHL	liquids	voiced stop	<190Hz	2	150	–0.1	
/ga/	HLL	voiced stop	unvoiced stop		3	130	–0.6	
/to/	LLL	unvoiced stop	vowel		4	105	–0.2	
/u/	LL–	no consonant	silence		5	90	–0.1	

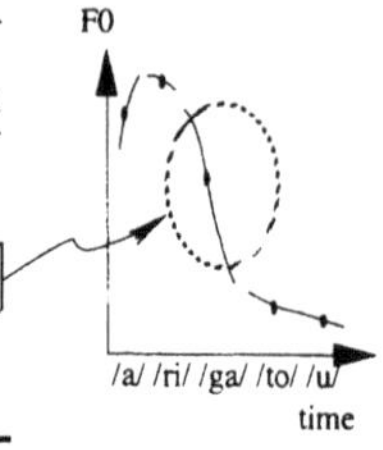

Figure 2.7. Example of syllable based F_0 units (SBU). M. Abe, H. Sato, "Two-stage F_0 control model using syllable based F_0 unit," *Proc. ICASSP92*, Vol. **2**, pp, 53–56 (© 1992 IEEE).

relatively high F_0 value, and the further a syllable is from the beginning of a phrase, the smaller its F_0 value.

The F_0 value for each syllable position is represented by average F_0 values, and difference F_0 values within a syllable. The model to determine one of the two values is called the syllable based F_0 unit (SBU).

Each syllable has a pair of SBUs, one for average F_0 and one for difference F_0, and sets of paired SBUs are generated according to syllable positions. In total, 10 SBUs are generated for 5 syllable positions, and the 5-th SBU deals with all syllables after the 5th syllable. The average F_0 values and difference F_0 values are determined by the attributes of following and current syllables. Figure 2.7 shows an example of typical attributes. Tone pattern is a high-low tone attribute for preceding, current, following syllables. The maximum F_0 value is assigned only one value for a phrase. The attribute called syllable position is used to estimate the F_0 values of syllables located later than the 5th syllable position. The attribute weights are obtained by the quantification theory (type one) which is a statistical approach [Hayashi-50].

An experiment was carried out using five hundred and three phonetically balanced sentences (total of 12000 syllables uttered by a professional announcer. Table 2.3 shows the results. The multiple correlation coefficient indicates the model's preciseness; if it is 1.0, the model is perfect, and the partial correlation coefficient shows how important the factor is in estimating values. In terms of average F_0 model, the measured multiple correlation coefficients indicate that the SBU model works well. Judging from the partial correlation coefficients, syllable position does not alter the impact of the control factors, i.e., maximum F_0 values and tone patterns are always the most and the second most important factors respectively. In terms of the difference F_0 models, the most interesting point is that every model has specific characteristics determined by syllable position. This is especially true for the first syllable, i.e., the maximum F_0 value is most important followed by tone pattern. Moreover, consonant class strongly determines F_0 change in the first syllable.

F_0 Contour Generation Rules

Early findings

Even if a sentence is spoken without any emphasis, the magnitudes of the accent components in a breath group are not the same because of various factors. Therefore, the magnitude of the accent component must be appropriately assigned to synthesize natural sounding speech. It is now commonly accepted that sentence structure, accent type, and phrase length in mora influence the magnitude of the accent component [Hakoda-76] [Hirose-79] [Hakoda-80] [Hirose-81] [Hirose-84] [Fujisaki-84]. These results were obtained by creating well-formed test sentences and analyzing their F_0 contours; and the analysis results were confirmed by listening tests. The influence of sentence structure is shown in Figure 2.8 [Hakoda-80]. When, as in Figure 2.8(A), the preceding phrase directly modifies the following one; i.e., adjacent phrases are strongly connected, the magnitude of the following accent component is suppressed. This kind of magnitude assignment for adjacent phrases is called "tight connection type." On the contrary, as in Figure 2.8 (B), when the preceding phrase modifies a distant one; i.e., adjacent phrases are weakly connected, the magnitude of the following accent component is increased. This kind of magnitude assignment for adjacent phrases is called "loose connection type." Using the hat pattern model, Figure 2.9 shows a rough sketch of F_0 contour according to strength of adjacent phrase connection and accent type. In terms of the influence of phrase

Table 2.3. Partial and multiple correlation coefficients of SBUs. M. Abe, H. Sato, "Two-stage F0 control model using syllable based F_0 unit," *Proc. ICASSP92*, Vol. 2 pp, 53–56 (© 1992 IEEE).

| SBU models | Tone Pattern | Partial Correlation Coefficients | | Max F_0 in Current Phase | Position Within a Phase | Multiple Correlation Coefficient | RMS error (Hz) | |
| | | Consonant Class | | | | | Close | Open |
		Current Syllable	Following Syllable					
1st syllable	0.553/0.215	0.287/0.399	0.403/0.452	0.836/0.480	—/—	0.876/0.654	116.6/0.56	15.8/0.54
2nd syllable	0.589/0.629	0.122/0.314	0.138/0.266	0.948/0.107	—/—	0.952/0.661	12.6/0.55	13.0/0.52
3rd syllable	0.762/0.361	0.203/0.235	0.141/0.303	0.877/0.305	—/—	0.909/0.557	18.4/0.42	20.5/0.41
4th syllable	0.703/0.424	0.121/0.328	0.166/0.257	0.798/0.333	—/—	0.852/0.590	20.9/0.42	21.7/0.40
other syllables	0.543/0.407	0.131/0.279	0.237/0.214	0.607/0.182	0.293/0.091	0.786/0.523	21.7/0.42	22.0/0.39

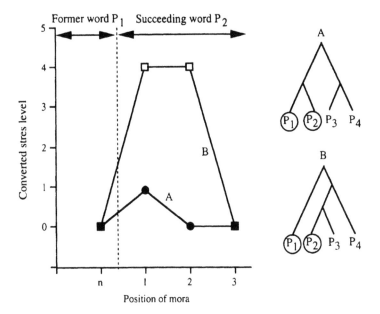

Figure 2.8. Examples of sentence structure. K. Hakoda, H. Sato, "Prosodic rules in connected speech synthesis," *Systems, Computers, Controls*, Vol. **11**, No. 5, pp. 28–37 (1980).

length in mora, the longer the phrase length, the larger the magnitude of its accent component.

Taking strength of adjacent phrase connection and phrase length in a mora into account, a rule to assign strong or weak connection type to each phrase boundary was proposed. In the rule, strength of adjacent phrase connection is qualified by "separation degree" that is the number of phrases from the current phrase to the phrase that is modified by the current phrase. The degree is assigned to every phrase boundary. Figure 2.10 shows an example of the separation degree. According to the observation of 122 speech samples, it was found that the product of the separation degree and the phrase length in a mora well estimates the magnitude assignment of strong or weak connection type. Figure 2.11 shows experimental results.

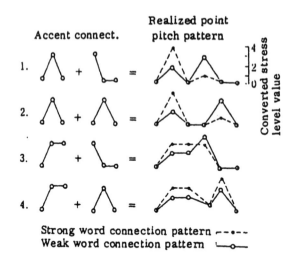

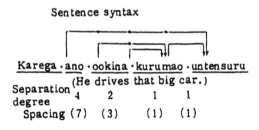

Figure 2.9. Point pitch concatenation pattern. K. Hakoda, H. Sato, "Prosodic rules in connected speech synthesis," *Systems, Computers, Controls*, Vol. **11**, No. 5, pp. 28–37 (1980).

Sentence syntax

Karega · ano · ookina · kurumao · untensuru
(He drives that big car.)
Separation degree   4   2   1   1
Spacing (7) (3) (1) (1)

Figure 2.10. Relation between sentence structure and separation degree. K. Hakoda, H. Sato, "Prosodic rules in connected speech synthesis," *Systems, Computers, Controls*, Vol. **11**, No. 5, pp. 28–37 (1980).

Quantification theory (type one)

This is a kind of linear regression model; it formulates the relationship between categorical and numerical values as follows; [Hayashi-1950]

$$\hat{y}_i = \bar{y} + \sum_f \sum_c x_{fc} \delta_{fc}(i)$$

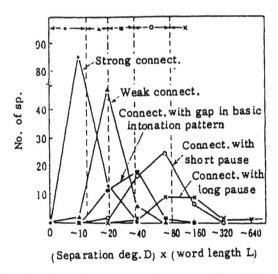

Figure 2.11. Relation between point pitch concatenation pattern and depth of phrase boundaries. K. Hakoda, H. Sato, "Prosodic rules in connected speech synthesis," *Systems, Computers, Controls*, Vol. **11**, No. 5, pp. 28–37 (1980).

where, $\hat{y}_i$ is the predicated value of the i-th sample, $\bar{y}$ is the mean value of all samples and $\delta_{fc}(i)$ is the characteristic function

$$\delta_{fc}(i) = \begin{cases} 1, & \text{if sample } i \text{ falls in category } c \text{ of factor } f \\ 0, & \text{otherwise.} \end{cases}$$

x_{fc} obtained by minimizing the following value.

$$(\hat{y}_i - y_i)^2$$

x_{fc} is the regression coefficient of the usual linear regression model. We call this value the "score" of a category.

An F_0 contour generation model was proposed using this algorithm,. The model is divided into two stages: global model and local model. The global model assigns the maximum F_0 values for minor phrases that constitute the sentence, and the local model assigns detail parameters within each minor phrase. The local model consists of a syllable based model, and experiment results have already been mentioned (p.19). Here we will look at the global model [Abe-92]. Factors are boundary type, accent type, syllable number in current phrase, and part of speech.

Table 2.4. Attributes of a minor phrase. M. Abe, H. Sato, "Two-stage F_0 control model using syllable based F_0 unit," *Proc.ICASSP92*, Vol. 2 pp, 53—56 (© 1992 IEEE).

Boundary Type	Accent Type	Part of Speech
top of a sentence,	with downfall,	interjection,
end of a sentence,	without downfall	adjective,
preceded by pause,		numerical,
followed by pause,		adverb, noun,
tight connection,		pronoun, verb,
loose connection		auxiliary verb

Table 2.4 shows categories of the factors. In Table 2.4, "Tight connection" and "loose connection" are related to sentence structure. "Tight connection" means that a current minor phrase directly modifies the following phrase or is directly modified by the preceding phrase, and "loose connection" means all other modification relationships. 503 phonetically balanced sentences (a total of 2890 minor phrases) were used in the experiments. These sentences were extracted from newspapers and magazines. Table 2.5 shows the experimental results. The multiple correlation coefficient indicates the model's preciseness; if it is 1.0, the model is perfect. Therefore, the average measured value of 0.843 indicates that the global model has good performance. The partial correlation coefficient means how important a factor is in estimating values. The experimental results show that boundary types and syllable numbers are the most important factors. Figure 2.12 shows category scores of boundary factors. The category scores mean the offset from an average value. Tight connection always emphasizes the height difference between the current phrase and preceding or following phrase, but the loose connection works conversely. In terms of syllable number effects, the longer a phrase is, the higher the maximum F_0 value of the phrase becomes. Accent types and part of speech have secondary importance. Accent type without downfall gives higher F_0 than accent type with downfall. These results agree with the early finding, but it is very important that the early finding were confirmed for not only well and intentionally constructed sentences, but also sentences used in newspapers and magazines. The effects of part of speech is a new finding. Figure 2.13 shows scores of the category for part of speech.

Table 2.5. Partial and multiple correlation coefficients of a global model. M. Abe, H. Sato, "Two-stage F_0 control model using syllable based F_0 unit," *Proc.ICASSP92*, Vol. 2 pp. 53–56 (© 1992 IEEE)

| Boundary Type | | Partial Correlation Coefficients | | | | | Multiple Correlation Coefficient | RMS Error (Hz) |
| | | | Accent Type | | Syllable Number in Current Phrase | Part of Speech | | |
Preceding	Following	Preceding Phase	Current Phrase	Following Phrase				
0.520	0.349	0.173	0.330	0.022	0.415	0.274	0.84	23.00

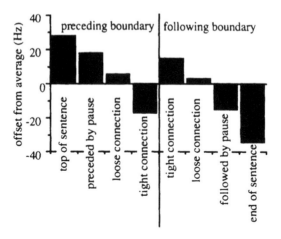

Figure 2.12. Effects of boundary. M. Abe, H. Sato, "Two-stage F_0 control model using syllable based F_0 unit," *Proc. ICASSP92*, Vol. **2**, pp, 53–56 (© 1992 IEEE).

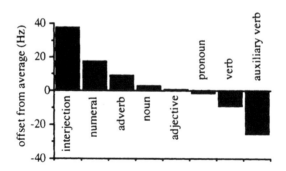

Figure 2.13. Effects of part of speech. M. Abe, H. Sato, "Two-stage F_0 control model using syllable based F_0 unit," *Proc. ICASSP92*, Vol. **2**, pp, 53–56 (© 1992 IEEE).

Tree-based approach

A decision tree [Quinlan-86] is one approach to classification problems. Once a decision tree is generated, unknowns are dropped into the tree at the top (root node), at each node various tests are applied to them to chose one branch or another unit and they finally reach the bottom (leaf) associated with a particular category. The approach is also applicable to assign numerical values instead of categories. Here, we will look at two kinds of algorithms based on the tree-based approach: decision tree and SBR-tree (Single Best Rule tree) [Indurkhya-91].

The tree-based approach has the following advantages over other statistical methods such as quantification theory and neural networks.

(1) Easy introduction of heuristics in developing rules.
(2) Easy integration of different types of features such as numerical, categorical, and binary.
(3) Easy interpretation of generated rules.

The decision tree generation algorithm is as follows:

(1) Assign all training data to the root node.
(2) Split the data into categories in terms of a feature associated with data.
(3) Repeat (2) for all possible features and select a feature that gives minimum error and decompose a node into descendant nodes using the feature.
(4) Repeat (2) and (3) recursively at new nodes till all the nodes satisfy the terminating condition.

In a decision tree, because many and various tests are generated, it is difficult to obtain knowledge by interpreting the tests and there is no guarantee that the rules are not redundant. The SBR-tree was proposed to solve these problems. The SBR-tree generation algorithm is as follows:

(1) Generate a decision tree using all training data.
(2) Based on a criteria, select a best path from the root to a leaf. The selected set of tests are called the single best rule. Examples of the criteria are the shortest rule, the rule covering the most data and so on.
(3) Remove redundant tests from the Single Best Rule (SBR) if any and generate a new set of tests; i.e., when a test is removed from the SBR and the remaining tests cause no significant drop in performance, the test is removed.
(4) Remove data covered by the new set of tests from the training data, which results in constructing a new set of data.

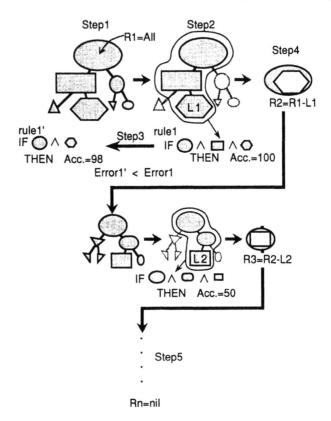

Figure 2.14. An algorithm of SBR-tree. M. Tanaka, Y. Nomura, Y. Yamashita, R. Mizoguchi, "Automatic generation of speech synthesis rules for accent components based on decision tree (in Japanese)," *Technical report of IEICE,* SP93–59, pp. 15–22 (1993).

(5) Generate a decision tree using the new set of data.
(6) Repeat steps (2), (3), (4), and (5) until the set of rules covers all data.

Figure 2.14 shows a rough sketch of the SBR-tree algorithm.

The tree-based algorithm was used to estimate the magnitudes of the accent component in the hat pattern model [Tanaka-93]. Two hundred long noun phrases (895 bunsetsu) from newspapers were spoken by a professional male speaker. Based on the hat pattern model, basic

```
rule1:                    [ COMP=yes ] -> 80.99  (250)
rule2:      [ LAST=yes, AUXV=yes ] -> 31.89  (19)
rule3:                  [ LAST=yes ] -> 52.72  (169)
rule4:                   [ VNN=yes] -> 95.70  (10)
rule5:                  [ AUXN=yes ] -> 49.60  (5)
rule6:      [ MOD2=yes, AUXV=no ] -> 77.05  (317)
rule7:                   [ WOV=yes ] -> 39.67  (3)
rule8:      [ AUXV=no, NIV=yes ] -> 42.00  (2)
rule9:                   [ NNO=yes ] -> 74.62  (24)
rule10:                  [ AUXV=no ] -> 63.26  (19)
rule11:                  [ NIV=yes ] -> 35.00  (4)
rule12:     [ MOD2=yes, HLOW=no ] -> 50.50  (10)
rule13:                 [ MOD2=yes ] -> 40.00  (9)
rule14:                        [ ] -> 32.00  (6)
```

Figure 2.15. A rule set generated by the SBR-tree algorithm. Y. Yamashita, M. Tanaka, Y. Amako, Y. Nomura, Y. Ohta, A. Kitoh, O. Kakusho, R. Mizoguchi, "Tree-based approaches to automatic generation of speech synthesis rules for prosodic parameters," *IEICE Trans. Inf. & Syst.*, Vol. **E76-A**, No. 11, pp. 1934–1941 (1993).

intonation patterns and accent component patterns were extracted by visual inspection of the F_0 contours. It has been reported that 200 long noun phrases are enough to generate rules. Table 2.6 shows experimental results. It is clearly observed that the SBR-tree can provide a performance comparable with that of a decision tree with a smaller number of rules. Figure 2.15 shows an example of the rules so generated. The obtained rules mostly agree with the early findings. Some rules related to part of speech such as auxiliary verb and noun are new.

Multiple Split Regression

The motivation for proposing multiple split regression was the weakness of the conventional statistical approaches. Strictly speaking, it is not guaranteed that factors employed in prosody control rules do not interact

26 Key Technologies for Text-to-Speech

Table 2.6. Performance of rule sets with the reduced features. Y. Yamashita, M. Tanaka, Y. Amako, Y. Nomura, Y. Ohta, A. Kitoh, O. Kakusho, R. Mizoguchi, "Tree-based approaches to automatic generation of speech synthesis rules for prosodic parameters," *IEICE Trans. Inf. & Syst.*, Vol. E76—A, No. 11 (1993).

Method	Error (Hz)	Leave/Rules
Decision tree	14.2	49.3
SBR-tee	15.1	13.1

with each other to some extent. However, in the linear regression formulation, the factors are assumed to be independent. The tree-based approach deals with the non-linear dependence of factors by splitting the factor space. However, once a space is split into subspaces, the tree-based approach can not deal with factor interaction or dependency over the subspaces. Multiple split regression combines the advantages of both approaches; it can deal with the non-linear dependence of factors as in the tree-based approach and can share parameters among split subspaces as in linear regression formulation. This decreases the number of free parameters. Therefore, the algorithm makes it possible to generate compact and reliable rules using a limited amount of training data.

Multiple split regression algorithm [Iwahashi-93a]

MSR predicts a value of sample i as the sum of the node's values where nodes are selected according to factors assigned to the sample i as follows. The nodes constitute a path from root to leaf.

$$y_i = \sum_{j=1}^{m} a_j \delta_i(j) + \varepsilon$$

Here, m represents the number of nodes in a tree, a_j represents the value of model parameter at each node. ε is assumed to have normal distribution with mean 0. $\delta_i(j)$ is a binary variable:

$$\delta_i(j) = \begin{cases} 1, & \text{if sample } i \text{ satisfies condition of node } j \\ 0, & \text{otherwise.} \end{cases}$$

A condition for node j is represented as products and sums of category sets. In MSR, the number of free parameters is equal to or less than m because the parameters are shared. Using least squared error criteria, estimation values of $\hat{a}_1, \hat{a}_2...\hat{a}_m$ for $a_1, a_2...a_m$ can be obtained by the solution of the following normal equations:

$$\sum_{j=1}^{m} f(u,j)a_j = \sum_{i=1}^{n} y_i\delta_i \quad u = 1,2,\text{K},m.$$

Here, n represents the number of samples, and coefficients $f(u, v)$ represent the number of samples which respond to both node u and node v simultaneously:

$$f(u,v) = \sum_{i=1}^{n} \delta_i(u)\delta_j(v).$$

The algorithm was applied to estimate the magnitudes of the phrase command and the accent command in the Fujisaki model [Hirai-94]. The experiment used 200 sentences spoken by a professional announcer. The numbers of phrase and accent commands of the sentences were 739 and 1193 respectively. In analyzing phrase command, five factors were employed; i.e., the number of morae in preceding, current and succeeding phrases, and the existence of commas at preceding and succeeding boundaries. Because splitting occurred at rood node 9 times out of the total (16 times), control factors seem to be almost independent. Figure 2.16 shows the third split from the beginning. The results indicate the importance of the preceding phrase length. In analyzing accent command, 16 factors were employed; i.e., the number of morae, the accent type and the part of speech of the preceding, current and following phrases, the separation degree (refer p.18) of preceding and current phrases, the existence of commas at preceding and following boundaries, the serial position and relative position of the minor phrase, and the number of preceding accented phrases. The rules so generated indicate that the accent type of current phrase is the most important factor. In terms of total performance, estimation error was 2.91 semitones; i.e. 23.2 Hz around the 150 Hz range.

Database template model [Mochida-93]

This approach directly uses the speech database contents; i.e., the F_0 contour is generated directly from F_0 contours in a speech database; if no suitable F_0 contour exists, the F_0 contour closest to the desired F_0 contour is selected and is modified as little as possible. One advantage of this approach is that there is no need for F_0 contour modeling to preserve natural F_0 contours. Because it is difficult to store all possible F_0 contours in a database, a key point is how to appropriately generate the F_0 contours that are not in the database.

28

Key Technologies for Text-to-Speech

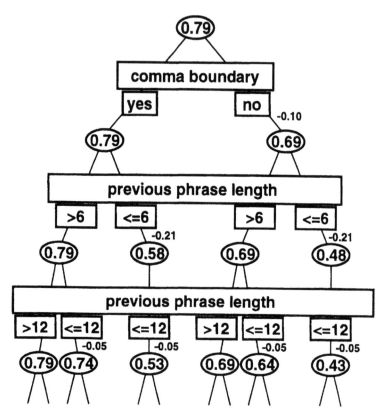

Figure 2.16. Splitting for phrase command. T. Hirai, N. Iwahashi, N. Higuchi, Y. Sagisaka, "Auto classification of F_0 control commands using statistical analysis," *Technical report of IEICE*, SP94–12, pp. 47–54 (1994).

An algorithm to generate F_0 contours in two steps was based on this idea. In the first step, the stored sentence closest to the target sentence is selected. The selected sentence is called the basic sentence. In the second step, in terms of number of mora and/or accent type, if the accent phrases of the basic sentence have different values from the target accent phrases, the closest accent phrase is selected as the basic accent phrase. Finally, the basic accent phrases are transplanted into the basic sentence. The procedure is as follows. The sentence structure of the basic sentence is completely identical with the sentence structure of the target sentence.

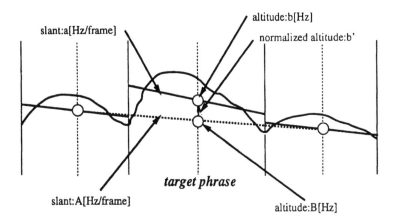

Figure 2.17. Approximation of F_0 contour.

Here, sentence structure simply indicates if a current accent phrase directly modifies the succeeding accent phrase or not. It was reported that this simple specification of sentence structure makes it possible to cover almost all sentences and introduces no performance degradation. If two or more basic sentences are found for the one target sentence, the most appropriate sentence is selected considering the similarities in terms of the number of morae and the accent type of the accent phrases. If accent phrases of the selected basic sentence have the exactly the same values as the target accent phrases in terms of the number of morae and accent type, the F_0 contour of the selected basic sentence is used to synthesize speech.

The second step is to select the basic accent phrase that has the same number of mora, the same type of accent, and the same sentence structure as the target accent phrase. To transplant the basic accent phrases into the basic sentence, the F_0 contour of the selected basic accent phrase is modified using the following a', b' parameters. Figure 2.17 shows the F_0 contour parameters of slant and altitude. The regression line for the F_0 contour is calculated phrase by phrase using the least squares method. The slant is assumed to be a, and the altitude of a center position to be b. The regression lines of the F_0 contour of the adjacent accent phrases of the target phrase are similarly calculated. A expresses the slant of the line that connects the center points of each accent-phrase. B expresses

Table 2.7. Normalization of F_0 contour for basic accent phrase.

	Accent Type	Number of Mora	Normalize to Basic Sentence	
			a'	b'
case1	0	X	Do	Do
case2	X	0	Not	Do
case3	X	X	Not	Do

(0: same, X: different)

the altitude of the line. The values (a,b) are normalized to (a', b') by the following equations.

$$a' = \frac{a - A}{1 + aA}$$

$$b' = b - B$$

Because it is necessary to preserve the transition of b' in the basic sentence, the value of b' of the basic accent phrase should be changed to match the value of b' in the basic sentence. On the other hand, because a' is strongly related to accent type, the value of a' of basic accent phrase should be preserved. Therefore, normalization is performed as shown in Table 2.7.

2.2.2. Phoneme Duration

Overview

Together with fundamental frequency, phoneme duration must be appropriately controlled to synthesize natural sounding speech and intensive studies have been done [Sato-77] [Higuchi-80] [Sagisaka-80] [Higuchi-81] [Sagisaka-81]. A framework of phoneme duration control has been developed in the same way as for fundamental frequency control; i.e., from hand made rules using a small amount of speech data to statistically optimized rules using large speech databases. Because preliminary research results are again important in the design of existing statistical models, we will look at the early findings.

Phonemes have intrinsic duration. For example, vowels /a/, /e/ and /o/ have relative longer duration than /i/ and /u/, voiceless fricatives /s/ and /ʃ/ have relative long duration, and liquid /r/ has relative short duration. Phoneme duration is influenced by neighboring phonemes. Figure 2.18 shows average phoneme duration of successive CV (consonants and vowel). As shown in Figure 2.18, average consonant duration is almost equal whatever vowel follows, but average vowel duration differs with the type of preceding consonant. It is interesting to note that vowel duration seems to change so as to keep the average phoneme length roughly the same. From this point of view, Japanese is often referred to as a mora-timed language; i.e., mora duration tends to be kept almost constant or if preceding consonant has short duration, the duration of following vowel is lengthened to compensate. Figures 2.19 and 2.20 show average consonant duration against average duration of succeeding vowels and against average duration of preceding vowels respectively [Sagisaka-80]. It is clearly shown that compensation is stronger in CV context than VC context. Phoneme duration is also affected by mora number in a breath group. Figure 2.21 shows the relationship between mora number in a breath group and average mora duration [Sagisaka-81]. It is observed that mora duration decreases according to the number of mora.

Statistical Approach

Quantification theory (refer p.18) was used in an experiment [Kaiki-92] to estimate vowel duration considering several factors such as syntactic, semantic, and segmental factors. The factors were: (1) the mora count of word, accent group, breath group and sentence in which the current vowel is included; (2) position of the current vowel in each group (initial, medial, or final); (3) part of speech (24 categories) to which the current vowel belongs; (4) neighboring phonemes (two preceding and two following); (5) gemination of preceding and following consonants; (6) accentedness; and (7) vowel category. A set of phonetically balanced 503 sentences was used. Four professional narrators (3 male and 1 female) spoken the set of sentences, and a model was generated for each speaker.

Table 2.8 shows the experimental results. It is clear that preceding phonemes, following phonemes and current vowel categories have the largest influence. This agrees with the early findings. Figure 2.22 shows the effect of the mora count of the utterance group. The more morae an utterance group has, the shorter the vowel duration is. This tendency is observed in sentences, breath groups and phrases, but not in words. Figure 2.23 shows the effect of position in the utterance group. Sentence-

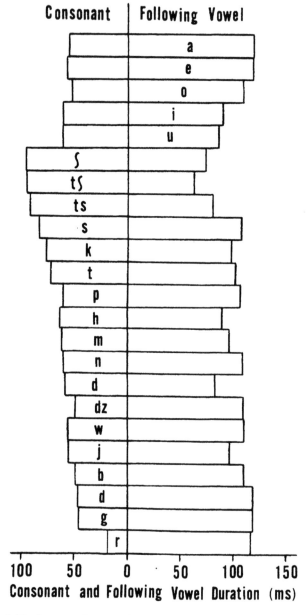

Figure 2.18. Average phoneme duration in word utterance. Y. Sagisaka, "A study of prosodic prameter control for speech synthesis," Phd. thesis (1986).

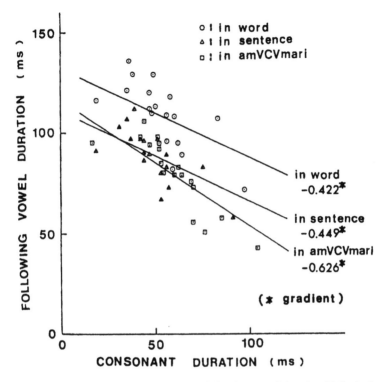

Figure 2.19. Consonant duration against following vowel duration. Y. Sagisaka, Y. Tohkura, "Characteristics of segmental durations in connected speech (in Japanese)," *Trans. of the committee on speech research, Acoust. Soc, Jpn.*, S80–34, pp. 267–273 (1980).

final shorting and phrase final lengthening are observed for all speakers. Figure 2.24 shows the effect of part of speech. Though the range of duration variation is small, the following characteristics were found. Quantifiers and proper nouns are lengthened more than normal nouns, and most particles and auxiliary verbs are shortened.

To generate a more compact duration control model, based on the above results, the number of factors was reduced to: (1) the mora count utterance group; (2) position in the utterance group; (3) part of speech; (4) neighboring phonemes; (5) gemination of adjacent consonants; (6) vowel category. The experimental results are shown in Table 2.9. Average prediction error was 15.84 msec.

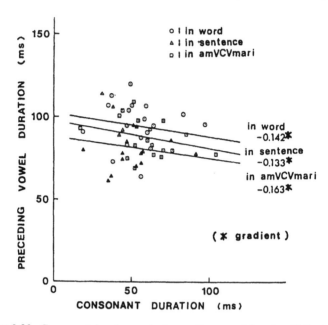

Figure 2.20. Consonant duration against preceding vowel duration. Y. Sagisaka, Y. Tohkura, "Characteristics of segmental durations in connected speech (in Japanese)," *Trans. of the committee on speech research, Acoust. Soc, Jpn.*, S80–34, pp. 267–273 (1980).

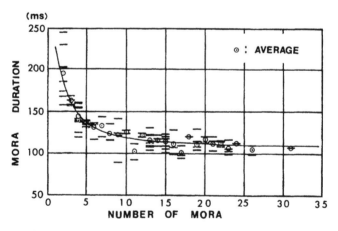

Figure 2.21. Relationship between mora number in a phrase and mora duration. Y. Sagisaka, Y. Tohkura, "Rule of segmental durations using statistical features of segment (in Japanese)," *Trans. of the committee on speech research, Acoust. Soc, Jpn.*, S80–72, pp. 561–568 (1981).

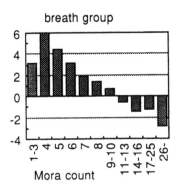

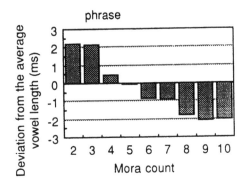

Figure 2.22. Relationship between mora number and mora duration. N. Kaiki, K. Takeda, Y. Sagisaka, "Linguistic properties in the control of segmental duration for speech synthesis," *Talking Machines: Theories, Models, and Designs*, G. Bailly, C. Benoit, and T. R. Sawalis (Editors), Elsevier Science Publishers, pp. 180–188 (1992).

2.2.3. Speech Power

Few studies have addressed speech power control [Mimura-91] [Itoh-93]. This is mainly because speech power has a smaller influence on synthesized speech quality than fundamental frequency and phoneme duration. In other words, synthesized speech quality was so low that controlling speech power offered little or no improvement. Synthesized speech quality is now high enough that a few studies on speech power control have been conducted.

A model for vowel power control was generated using the quantification theory [Mimura-91]. Fundamental frequency was added

Table 2.8. Partial correlation coefficients for vowel duration estimation. N. Kaiki, K. Takeda, Y. Sagisaka, "Linguistic properties in the control of segmental duration for speech synthesis," Talking Machines: Theories, Models, and Designs, G. Bailly, C. Benoit, and T. R. Sawalis (Editors), Elsevier Science Publishers, pp. 180—188 (1992).

	Factors	Speakers			
		A	B	C	F
Current Vowel		0.475	0.470	0.364	0.450
neighboring phonemes	preceding	0.564	0.592	0.488	0.592
	following	0.483	0.409	0.416	0.483
	second preceding	0.210	0.336	0.167	0.221
	second following	0.233	0.225	0.197	0.176
gemination	preceding	0.007	0.045	0.020	0.042
	following	0.130	0.036	0.132	0.035
utterance group position	sentence	0.428	0.524	0.323	0.282
	breath	0.218	0.406	0.486	0.121
	phrase	0.170	0.114	0.157	0.075
	word	0.012	0.013	0.022	0.062
utterance group length	sentence	0.069	0.083	0.068	0.055
	breath	0.146	0.119	0.139	0.143
	phrase	0.085	0.091	0.085	0.055
	word	0.042	0.036	0.060	0.073
accentedness		0.015	0.003	0.018	0.006
part of speech		0.115	0.166	0.171	0.133
multiple correlation coefficient		0.798	0.863	0.796	0.769

Table 2.9. Prediction errors of vowel duration. N. Kaiki, K. Takeda, Y. Sagisaka, "Linguistic properties in the control of segmental duration for speech synthesis," Talking Machines: Theories, Models, and Designs, G. Bailly, C. Benoit, and T. R. Sawalis (Editors), Elsevier Science Publishers, pp. 180–188 (1992).

Data	Speakers				Mean
	A	B	C	F	
Analysis vowels	14.58ms (19.2%)	14.90ms (18.1%)	16.38ms (22.7%)	15.32ms (18.4%)	15.30ms (19.6%)
Test vowels	14.87ms (19.6%)	15.47ms (18.8%)	17.08ms (23.7%)	15.92ms (19.2%)	15.84ms (19.9%)

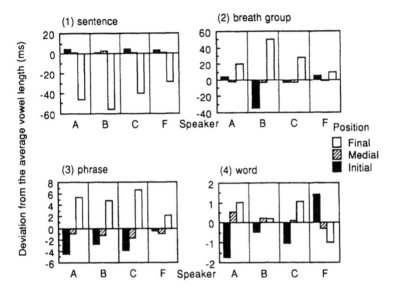

Figure 2.23. Vowel duration for utterance group position. N. Kaiki, K. Takeda, Y. Sagisaka, "Linguistic properties in the control of segmental duration for speech synthesis," *Talking Machines: Theories, Models, and Designs*, G. Bailly, C. Benoit, and T. R. Sawalis (Editors), Elsevier Science Publishers, pp. 180– 188 (1992).

to the control factors used in the duration control model described in the previous section. To generate the model, a set of phonetically balanced 503 sentences containing 11,384 vowel segments was used. Table 2.10 shows the multiple correlation coefficient and partial correlation coefficients. The results indicate that the important factors are fundamental frequency, preceding phoneme, succeeding phoneme and position in the utterance group. Figures 2.25, 2.26 and 2.27 show the category score of fundamental frequency, position in sentence, and the preceding phoneme respectively. It is observed that the higher the fundamental frequency, the larger the speech power, and the closer the vowels are located to the end of the sentence, the smaller the vowel power. If a vowel is preceded by a vowel, its power is increased. On the other hand, if a vowel is preceded by a voiceless consonant, its power is decreased.

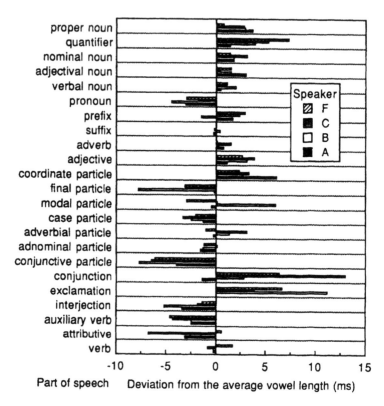

Figure 2.24. The difference in vowel duration by part of speech. N. Kaiki, K. Takeda, Y. Sagisaka, "Linguistic properties in the control of segmental duration for speech synthesis," *Talking Machines: Theories, Models, and Designs*, G. Bailly, C. Benoit, and T. R. Sawalis (Editors), Elsevier Science Publishers, pp. 180–188 (1992).

2.3. Speech Synthesis Units

Phonemic information is an essential attribute of speech. Because it is hard to develop rules that can generate all parameters related to phonemic information, almost all speech synthesis-by-rule systems store a set of speech segments uttered by human beings and use them as basic parameters in synthesizing speech. The speech segments are called speech synthesis units. The parameter sequences of phonemic

Table 2.10. Partial correlation coefficients for segmental phoneme amplitude estimation. K. Mimura, N. Kaiki, Y. Sagisaka, "Statistically derived rules for amplitude and duration control in Japanese speech synthesis," *Proc.Korea-Japan joint symposium on acoustics,* pp. 151–156 (1991).

Parameters		Correlation Coefficients	
	Category	Vowel Amplitude	Vowel Duration
current pheneme		0.420	0.475
neighboring phoneme	preceding	0.579	0.564
	following	0.569	0.483
	pre-adjacent	0.160	0.210
	post-adjacent	0.087	0.233
gemination	preceding	0.046	0.007
	following	0.017	0.130
position	sentence	0.443	0.428
	breath group	0.253	0.218
	phrase	0.039	0.170
	word	0.056	0.012
mora count	sentence	0.086	0.069
	breath group	0.136	0.146
	phrase	0.052	0.085
	word	0.032	0.042
accentedness		0.021	0.015
part-of-speech		0.146	0.115
fundamental frequency		0.797	—
multiple correlation coefficent		0.925	0.798

information to be synthesized are generated by concatenating synthesis units. For example, /arigatou/ is divide into synthesis units such as /a/ , /r/, /i/, /g/, /a/, /t/, /o/, /u/ or /a/, /ri/, /ga/, /to/, /u/ and so on. From a requirement of the speech synthesis-by rule; i.e., the system have to synthesize any kinds of phoneme sequences in a specific language, the synthesis units must cover all possible phoneme sequences. Therefore, synthesis unit length is usually shorter than a word. On the other hand, to synthesize natural and highly intelligible speech, a synthesis unit should contain as much coarticulation detail as possible. From these points of view, several kinds of synthesis units have been proposed based on heuristic knowledge. In the first half of this section, we will look at these synthesis units. In terms of preserving coarticulation, the longer synthesis units generally have advantages. However, from a practical point of view; i.e., to develop a speech synthesis-by-rule system with reasonable cost, the longer the synthesis units are, the more units are

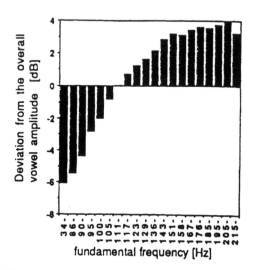

Figure 2.25. Vowel amplitude as a function of F_0. K. Mimura, N. Kaiki, Y. Sagisaka, "Statistically derived rules for amplitude and duration control in Japanese speech synthesis," *Proc. Korea-Japan joint symposium on acoustics*, pp. 151–156 (1991).

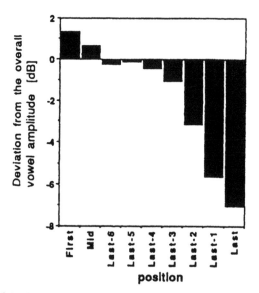

Figure 2.26. Vowel amplitude as a function of position in a sentence. K. Mimura, N. Kaiki, Y. Sagisaka, "Statistically derived rules for amplitude and duration control in Japanese speech synthesis," *Proc. Korea-Japan joint symposium on acoustics*, pp. 151–156 (1991).

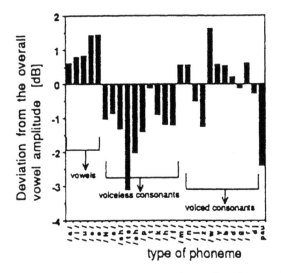

Figure 2.27. Vowel amplitude as a function of preceding phoneme. K. Mimura, N. Kaiki, Y. Sagisaka, "Statistically derived rules for amplitude and duration control in Japanese speech synthesis," *Proc. Korea-Japan joint symposium on acoustics*, pp. 151–156 (1991).

needed, more memory area is required, and more effort is needed to generate the synthesis units. Therefore, there is no precise solution to this problem, and we have to combine these units according to some conditions. Algorithms to design an optimal set of synthesis units or to obtain an optimal synthesis units sequence against the input text were recently proposed. The key points of these algorithms are (1) use of large scale speech databases, and (2) introduction of acoustic measure in determining synthesis units. The second part of this section explains some of these approaches.

2.3.1. Heuristically Designed Units

Phoneme

Because the phoneme is the minimum unit that can be clearly related to linguistic symbols and the number is rather small (from 20 to 30 in Japanese), it is a candidate for synthesis units. However, the simple phoneme unit is not popular in Japanese speech synthesis-by-rule, because the speech so synthesized has poor quality for the following reasons (1) it is hard to generate unit concatenation rules to represent the coarticulation phenomena that occurs between phoneme units; and (2) unit concatenation generally causes unnatural sound and unit concatenation occurs too frequently. A more sophisticated usage of phoneme units was proposed recently. As shown later, the algorithm proposed automatically generates a set of phoneme units. Another approach [Hakoda-95] is to generate phoneme units by considering the preceding phoneme and following phoneme, the so called tri-phone context; i.e., different /o/ units, for example, are prepared for use with /kos/ and /bos/. The tri-phone context makes it possible for the phoneme unit itself to include coarticulation phenomena, and the unit concatenation rules become simpler. Using the tri-phone context, about 15000 units are necessary to cover Japanese sentences.

CV (Consonant-Vowel) Unit

Because Japanese syllables basically consist of a consonant-vowel structure, CV units are suitable as Japanese synthesis unit [Tohkura-80]. The advantage of the CV unit is in preserving the coarticulation part between consonant and vowel. Only 100 to 130 CV units are needed to cover Japanese including words of foreign origin. A problem is how to concatenate CV units; i.e., the coarticulation part between the vowel of the preceding CV and the consonant of the following CV must be generated by rules.

VCV (Vowel-Consonant-Vowel) Unit

A VCV unit can preserve the coarticulation part between vowel and consonant and between consonant and vowel [Sato-78]. The motivation for creating the VCV unit was to preserve more coarticulation information than possible with the CV unit. Another advantage is that VCV units can be concatenated at vowel centers. Because vowels are relatively stationary in speech, the unit concatenation rules are much simpler than the rules for vowel-consonant concatenation in the case of CV units. There are about 800 units including VV units and CV units for the first syllable in a phrase.

CVC (Consonant-Vowel-Consonant) Unit

An advantage of the CVC unit is the unit concatenation during consonants [Sato-84]. Because vowels have more power than consonants, vowel quality determines total synthesized speech quality from the perceptual point of view. Therefore, inadequate unit concatenation within the vowel region causes more serious degradation than concatenation within the consonant region. Moreover, vowel characteristics are heavily influenced by surrounding phonemes. Therefore, the concatenation within the vowel region is more difficult than that in the consonant region. Another advantage is better preservation of the coarticulation region between consonant and vowel; i.e., there are more CVC combinations than VCV combinations because the number of consonants is more than double the number of vowels. Because the number of CVC units needed to cover all Japanese phoneme sequences is large, a practical system must use a limited number of CVC units. An analysis of a Japanese word dictionary showed that 1100 CVC units cover 90% of the phonemes sequences in the dictionary. The remaining phoneme sequences are covered by using CV, VC, and VV units. Consequently, the total number of synthesis units is about 1300.

2.3.2. Database Oriented Units

COC Method [Nakajima-88]

As an alternative to heuristically designed synthesis units, an algorithm was proposed that automatically generates synthesis units using acoustic measures. The algorithm is based on segment quantization techniques, and one unique point is that it generates a codebook so that each cluster has a phoneme label entry. Therefore, the proposed algorithm is referred to as Context Oriented Clustering (COC). Figure 2.28 shows the outline of COC-based speech synthesis. The method consists of two major phases: a synthesis unit generation phase and a speech synthesis phase. The synthesis unit generation phase is an off-line process that uses a speech database with phonological information (e.g. phonetic labels). Feature extraction analysis, such as LPC analysis, is performed on the database to obtain feature vector sequences. Using feature vector sequences with phonetic labels, a set of appropriate synthesis units is generated through the COC procedure. Figure 2.29 shows the simple cluster partitioning process in COC. Initially, each cluster consists of those segments that have the same phonetic labels. Each cluster is then split iteratively by the restriction of preceding or following contexts. In Figure 2.29, initial cluster W_1, which is a set of the segments with phonetic label /a/, is split into two clusters: W_{11}:/a/ preceded by /b/, and

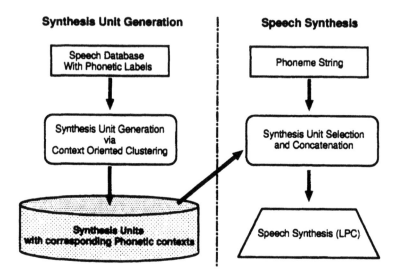

Figure 2.28. Outline of COC-based speech synthesis. S. Nakajima, H. Hamada, "Automatic generation of synthesis units based on context oriented clustering," *Proc. ICASSP88*, pp, 659–662 (© 1988 IEEE).

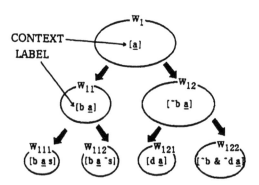

Figure 2.29. An example of cluster partitioning process in COC. S. Nakajima, H. Hamada, "Automatic generation of synthesis units based on context oriented clustering," *Proc. ICASSP88*, pp, 659–662 (© 1988 IEEE).

W12: /a/ preceded by any phoneme but /b/. These two clusters can be further split into W111 (/a/preceded by /b/ and followed by /s/) and W112 (/a/ preceded by /b/ and followed by any phoneme but /s/). This iterative process is continued until some terminal conditions are satisfied. As shown in Figure 2.29, each cluster has a symbolic representation, called a context label, identifying what phoneme the cluster represents and in what context the cluster is to be used.

In the on-line speech synthesis phase, given a phoneme string, the best synthesis units are selected on the basis of context matching score against the given phonetic context. By concatenating selected units, a sequence of feature parameter vectors is obtained. The parameter sequence is then passed through a speech synthesizer to generate speech waveforms.

Details of the COC algorithm are as follows:

Step 1: Let each initial cluster Wi consist of the segments with the same phonetic labels, compute each *cluster evaluation value* n_i. Let the initial clusters that have more than Nmin sample segments be the members of the search set y.

Step 2: Of all the clusters in y, find cluster W_k that has maximum ni and has more sample segments than N_{min}. If no such cluster exists, go to Step 6.

Step 3: For every context layer, find the new additional context term g* such that *split evaluation value* is maximum and that the segments that satisfy g* are more than N_{min}. In addition, g* must satisfy the *split criterion*. If no such context term exists, remove W_k from y, and go to Step 2.

Step 4: Split cluster W_k into two clusters of segments by g*: segments that satisfy context g* (W_{k1}) and all others (W_{k2}). Next, delete W_k from y and put W_{k1} and W_{k2} into y. (As an exception, if the segments which do not satisfy g* are less than N_{min}, W_k itself is taken as W_{k2}.)

Step 5: If the terminate condition is not satisfied, go to Step 2.

Step 6: The prototype of each final cluster (mostly the centroid matrix) is saved with its context label.

Examples of the *cluster evaluation value*, the *split evaluation value*, and *split criterion* are an inner-cluster variance normalized by the average length, the difference between the cluster values after splitting, and the average of the cluster evaluation values, respectively. For centroid calculation, each segment is converted to a fixed dimension L by using linear interpolation or resampling, and the L is not the same value for all clusters but is calculated for each cluster. Thus, each centroid matrix also retains duration information

Synthesis unit selection

In the speech synthesis phase, given a phoneme string, those synthesis units having context labels that have the best *context matching score* against a given context are selected. The *context matching score* of context label s against given context h is denoted by $M_d(s, h)$ and can be computed as follows. Let s be $[E_k,....,E_2, E_1, x, E'_1, E'_2,... E'_l]$ and h be $/p_n,..., p_2, p_1, x, q_1, q_2,... q_m/$. The context matching score can be expressed by the sum of *pre/post-context matching score*, as follows:

$$M_d(s,h) = \sum_{i=1}^{k} m(p_i, E_i) + \sum_{j=1}^{t} m(q_j, E'_j)$$

where $m(p, E)$ denotes the matching degree between phoneme p and context expression E, and can be defined as follows.

$$m(p,E) = 1, (E = p)$$

$$= \frac{1}{n}, (E = p \mid g_1 \mid g_2 \mid L \mid g_{n-1})$$

$$= \frac{1}{N_{ph} - n}, (E = \tilde{g}_1 \& \tilde{g}_2 \& K \& \tilde{g}_n \text{ and } g_1 \neq p \text{ for } 1,2,K, n)$$

$$= -\infty, \text{ (otherwise)}$$

where N_{ph} is the number of phonetic symbols, & is logical-and, | is logical-or, and ~ is complement.

An experiment was carried out using 432 words uttered by a male speaker. Speech was digitized at 12 kHz sampling and 16-th order LPC analysis was performed. N_{ph} was 43, and N_{min} was 5. In total, 627 units were generated. Figure 2.30 shows the top 10 categories of the COC units. The experiment showed that various contexts are automatically generated and the results coincide with the phonological facts.

Non-uniform Unit [Iwahashi-93b]

A unique point of the non-uniform unit is that synthesis units are selected

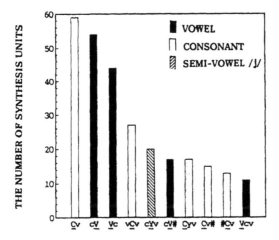

Figure 2.30. The number of classified synthesis units generated by COC. S. Nakajima, H. Hamada, "Automatic generation of synthesis units based on context oriented clustering," *Proc. ICASSP88*, pp, 659–662 (© 1988 IEEE).

based on not only phoneme information but also other factors including F_0, duration and so on. In all other units mentioned above, sets of synthesis units are generated in advance and once phoneme strings to be synthesized are given, a unique sequence of the synthesis unit are determined, so there is no way to take factors other than phoneme information into account when selecting synthesis units.

However, the non-uniform unit approach is different. This approach prepares a database in advance, such as many word and/or sentence utterances. Taking into account F_0, duration and synthesis unit concatenation condition, synthesis units are selected from the database to optimally match the phoneme strings to be synthesized. For example, phoneme sequence "$C_1V_1C_2V_2C_3V_3C_4V_4C_5V_5$" in the database might be used as "$V_1C_2V_2$", "V_2C_3", "$C_1V_1C_2$", or "$V_2C_3V_3$" and so on. Also, the selected synthesis units for synthesizing a sentence might consist of phoneme sequences of various length, such as CV, CVC, CVCV and so on, and it might be possible that even a word or phrase is selected as a synthesis unit. Because many factors influence synthesized speech quality, a key point in this approach is the algorithm used to locate the optimal synthesis units from within the given speech database.

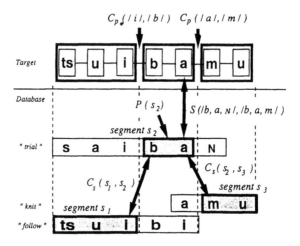

Figure 2.31. Four acoustic measures to evaluate spectral distortion. N. Iwahashi, N. Kaiki, Y. Sagisaka, "Speech segment selection for concatenative synthesis based on spectral distortion minimization," *IEICE Trans. Fundamentals*, Vol. E76-A, No. 11, pp. 1942–1948 (1993).

Based on an interpretation of heuristic knowledge, one algorithm defines the search criterion using four acoustic measurements; $S(e_s, e_t)$, $P(s)$, $C_p(/p_p/,/p_f/)$ and $C_s(s_p, s_f)$. Figure 2.31 shows an example of applying the four acoustic measures. $S(e_s, e_t)$ is spectrum distortion when the selected unit is used in a different phoneme context as shown in Figure 2.31; i.e., phoneme sequence /ba/ from /baN/ is used in /bam/ . $S(e_s, e_t)$ is calculated as the Euclidean distance between the centroids of the sets of time-normalized cepstral parameters of the segments in the two contexts. Only the immediately preceding and succeeding phonemes are considered as context. $P(s)$ indicates the prototypicality of segment s. The prototypicality is based on the fact that, even though a segment is extracted from the desirable context, the segment could have different characteristics from average characteristics because of other factors such as recording condition or mispronunciation and so on. In such a case, the selected segment should not be used as synthesis unit. $P(s)$ is calculated as the Euclidean distance between time-normalized cepstral parameters of segment s and the centroid of the set of segments from the context. $C_p(/p_p/,/p_f/)$ is a measure of how much the concatenation of two phonemes will degrade speech quality. The value is calculated by subtracting the mean of the spectral change at the boundaries of phonemes $/p_p/$ and $/p_f/$ from the maximum of these mean values. This measurement is based on experiment results indicating that concatenation at rapid spectrum changes is less audible than concatenation at stable spectrum changes. $C_s(s_p, s_f)$ is a measurement of discontinuity between segment s_p and segment s_f. This is calculated as the spectral distance between s_p and s_f around the concatenation point.

Synthesis unit selection based on the acoustic measurements is performed in three steps. At first, the search space is reduced by evaluating $P(s)$. Unusual or irregular speech segments are discarded from the database. The second and third steps deal with optimization at different levels; i.e., phonemic symbol level and speech segment level. In the second step, the target phoneme sequences (phoneme sequences to be synthesized) are divided into optimal phoneme sub-sequences using the following cost function.

$$f_2 = \sum_{\text{over sentence}} \left\{ S(e_s, e_t) + \alpha \times C_p(/p_p/, /p_f/) \right\}$$

This cost function is minimized by dynamic programming as follows:

1. Set initial value,

$$f_2(0) = 0$$

2. Repeat the following procedure for $i = 1, 2, ..., N_p$

$$f_2(i) = \min_{0 \le j \le i-1} \left\{ f_2(i-1-j) + d(i-j, i) \right\}$$

$$d(i-j, i) = \min_{seg(i-j, i, p_p, p_f)} \left\{ \begin{array}{l} S(/p_p, p_{i-j}, p_{i-j+1}/, /p_{i-j-1}, p_{i-j}, p_{i-j+1}/) \\ +S(/p_{i-1}, p_i, p_f/, /p_{i-1}, p_{i+1}/) \\ +\alpha \times C_p(/p_{i-j-1}, p_{i-j}) \end{array} \right\}$$

on condition that $seg(i-j, i, p_p, p_f)$ exists in the database.

3. Get the minimum value of f_2 by

$$\min f_2 = f_2(N_p)$$

where N_p and p_i represent the number of phonemes and the i-th phoneme in the target phoneme sequence, respectively. $seg(i-j, i, p_p, p_f)$ represents the segment in the phoneme sequence $/p_i, p_{i+1}, ..., p_j/$ extracted from the context preceded by $/p_p/$ and followed by $/p_i/$.

The process up to this point has determined the optimal phoneme sub-sequences. Next, the optimization moves from the phonemic symbol level to the acoustical speech segment level. In this step, the optimal speech segment sequence is finally selected from the speech segments belonging to the optimal phoneme sub-sequence. Automatic boundary adjustment according to the phonemic environment is carried out. The cost function f_3 for this step is as follows:

$$f_3 = \sum_{\text{over sentence}} C_s(s_p, s_f)$$

This cost function is minimized by dynamic programming as follows:

1. Set initial value,

$$f_3(0,0) = 0$$

2. Repeat the following procedure for $i = 1,2,...,N_s$

$$f_3(i,j) = \min_k \left\{ f_3(i-1,k) + C_s(s_{i-1,k}, s_{i,j}) \right\}$$

3. Get the minimum value of by f_3 following:

$$\min f_3 = \min_j f_3(N_s, j)$$

where N_s represents the number of phoneme sub-sequences determined for the target phoneme sequences. $s_{i,j}$ represents a j-th speech segment candidate for i-th phoneme sub-sequence. An optimal sequence of speech segments is chosen from the segments which minimize the cost.

2.4. Speech Modification Algorithms

2.4.1. Overview

In speech synthesis-by-rule based on synthesis unit concatenation, synthesis unit modification in terms of prosodic parameters such as fundamental frequency, phoneme duration and speech power, is necessary because the synthesis units uttered by human beings have, in general, different prosodic parameter values from those desired. To modify the prosodic parameters, many algorithms have been studied over the last two decades; most are based on based on source-filter models. In the models, "source" is a glottal waveform and "filter" is the vocal tract; i.e., the glottal waveform produced by vibration of the vocal cords is modulated by passing through a particular vocal tract shape to pick up the desired phonemic information. The model allows source and filter to be controlled independently. Therefore, the model is suitable for changing the prosodic parameters, especially fundamental frequency, of synthesis units. Based on the model, a formant synthesizer was introduced [Allen-87] followed by an LPC synthesizer [Markel-76]. To improve LPC synthesizer quality, extensive studies examined the use of the residual signal as the source [Sato-84] [Tacked-85]. More recently, however, algorithms based on waveform manipulation have become popular. The algorithm called PSOLA (Pitch Synchronously Overlap Addition) was originally proposed for French [Moulines-90]. The PSOLA algorithm has several advantages. First, PSOLA manipulates waveforms and does not perform source-filter separation. Because the source-filter model is an approximation, there is no guarantee that the separation works well if the fundamental frequency is changed. Therefore, even if the residual signal is employed to improve LPC synthesizer quality, the synthesized speech is sometimes rough. Waveform manipulation is free of these problems. Second, PSOLA is performed pitch synchronously. This makes it possible to easily change fundamental frequency in the time domain. Based on the idea of PSOLA, several algorithm were proposed in Japan [Kawai-93] [Mitome-94] [Arai-94] [Sakamoto-95] [Katae-95], and in fact most commercial Japanese text-to-speech systems employ PSOLA-like algorithms. Concerning the development of TTS systems, another advantage of waveform-based algorithms is that the amount of computation needed is small.

2.4.2. A Speech Modification Algorithm Without Source-filter Separation

PSOLA has a problem when a target F_0 value is quite different from the F_0 value of available synthesis units. Because PSOLA simply extracts speech signals by cutting out within a window, separation between the two adjacent pitch segments is incomplete. This is one reason why PSOLA introduces some noise when F_0 modification is large. An algorithm was proposed to overcome this problem [Takagi-93]. Like PSOLA, the algorithm is performed pitch-synchronously and does not require source-filter separation. A unique point of the algorithm is that pitch period is first shortened or lengthened in the waveform domain; next, the distortion introduced by the waveform modification is compensated in the spectrum domain.

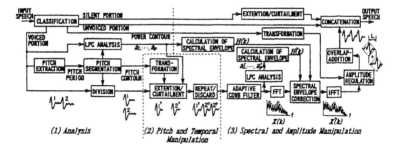

Figure 2.32. The block diagram of the speech modification algorithm. T. Takagi, E. Miyasaka, "A speech prosody conversion system with a high quality speech analysis-synthesis method," *Proc. Eurospeech93*, pp. 995–998 (1993).

Figure 2.32 shows a block diagram of the algorithm. An input speech waveform is classified into voiced, unvoiced and silent portions based on frame-synchronous analysis. The pitch periods of a voiced portion are extracted by peak picking method against low-pass filtered signal of the input waveform. Using the pitch periods, pitch segments are extracted from the input speech waveform. The start point of each pitch segment is zero-crossing point which locates the closest before to a point which gives the local maximum power. LPC coefficients $a_1, a_2,..., a_p$ are calculated for speech signal which is extracted using Hamming window centered at the middle point of the pitch segment. Here, we will look at how to change original F_0 value to a target F_0 value. Let k and k' denote the sampling points for original and the target pitch period, respectively. As shown Figure 2.33, to shorten the pitch period, the signals are truncated at the k'-th point, and to lengthen the pitch period, $x(m)$ is estimated from $m = k + 1$ to k' using the LPC coefficients $a_1, a_2,..., a_p$ as follows.

$$x(m) = \sum_{i=1}^{p} a_i x(m-i) e^{-r(m-k)},$$

where r is a damping factor which is positive; r depends on the peak level of the original pitch segment and the amount of lengthening. Temporal structure of a voiced portion can be manipulated by occasional repetition or discarding of individual pitch segments. To remove the discontinuity that occurs between the last point of a pitch segment and

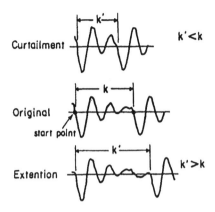

Figure 2.33. Extension or curtailment of a pitch period. T. Takagi, E. Miyasaka, "A speech prosody conversion system with a high quality speech analysis-synthesis method," *Proc. Eurospeech93*, pp. 995–998 (1993).

the start point of the succeeding pitch segment, the least square approximation method with cubic function is employed using several sampling points near the connecting point. This process modifies the pitch period, but some spectrum distortion is introduced. In the following, the spectrum distortion is compensated. The compensation is performed frame-synchronously. The sequence $x(m)$ is cut from the modified speech using an N-length Tukey window. N is an integer power of two. First, to reduce the perceptual distortion to some extent, adaptive comb filtering with K_p (the average pitch period within the frame) is performed as follows:

$$x'(m) = h_1 x(m - K_p) + h_0 x(m) + h_1 x(m + K_p),$$

where, $h_0 = 0.5 \times (1 + \alpha)$, $h_1 = 0.25 \times (1 - \alpha)$, and α is chosen between 1.0 and 0.01; smaller values are used as the amount of pitch manipulation increases. Next, the speech signal is extracted from $x'(m)$ using a Hamming window, and LPC coefficients a'_1, a'_2,..., a'_p and Fourier coefficients $X'(k)$ are calculated for the speech signal. The spectrum distortion is compensated using the following equation.

$$X''(k) = \frac{H(k)}{H'(k)} X'(k) \quad (k = 0,1,\text{K}, N-1)$$

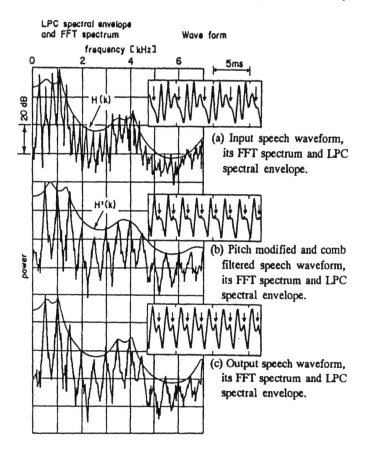

Figure 2.34. FFT spectra, LPC spectral envelope and waveforms in an example of uniform pitch shift. T. Takagi, E. Miyasaka, "A speech prosody conversion system with a high quality speech analysis-synthesis method," *Proc. Eurospeech93*, pp. 995–998 (1993).

where

$$H(k) = \frac{1}{\left|1 + \displaystyle\sum_{i=1}^{p} a_i e^{-j\frac{2\pi i k}{N}}\right|} \quad (k = 0,1,\mathrm{K}\,,N-1)$$

$$H'(k) = \frac{1}{\left|1 + \displaystyle\sum_{i=1}^{p} a_i' e^{-j\frac{2\pi i k}{N}}\right|} \quad (k = 0,1,\mathrm{K}\,,N-1)$$

The modified $X''(k)$ are transformed into the time domain sequence $x''(m)$ by FFT. A two-window-shift length sequence is extracted from $x''(m)$ using a Hamming window, and is overlap added to adjacent frame signals. Figure 2.34 shows an example of spectrum distortion compensation for a uniform pitch shift. According to listening tests for pitch shift modification from –1 to +1 octave, the algorithm can produce higher quality speech than the conventional LPC analysis-synthesis method.

CHAPTER 3

Evaluation Method of Text-to-Speech Synthesis

In recent years, commercial text-to-speech (TTS) products have become available. However, no current TTS can offer the best performance for all kinds of applications, so users have to select the most appropriate TTS for their particular need. Therefore, TTS system performance must be measured carefully. Because, as shown in Chapter 2, TTS is a highly integrated system of text analysis, phonetics, phonology and digital signal processing, total TTS performance can not be indicated by a single evaluation measure, but must be evaluated from various kinds of aspects. In this section, we will first look at examples of evaluation, and then show a proposed guideline for TTS evaluation.

3.1. Evaluation Experiments

Considering synthesized speech as a means of communication, intelligibility is one of the most important aspects. To evaluate intelligibility, word intelligibility is a good candidate, because a word is a basic unit for sentence understanding. Word intelligibility was investigated in terms of word attributes such as word length, word familiarity, initial phoneme, and the existence of similar words [Watanabe-88]. A set of words was selected so that each word attribute occurred with equal frequency in the set. Three TTS systems were evaluated together with natural speech and analysis-synthesis speech. Figure 3.1 shows the experimental results. The score is relative word intelligibility against the average word intelligibility in each kind of speech. It is clearly shown that word familiarity greatly influences intelligibility. An interesting point is that the difference among the 3 TTS systems is less than that in natural speech. The initial phoneme in a word also greatly influences word intelligibility. A word whose initial

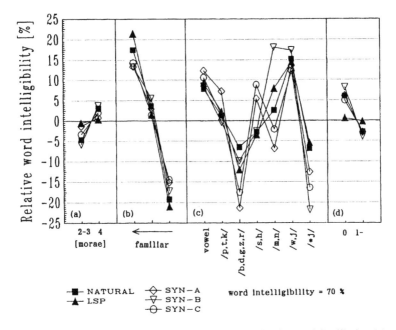

Figure 3.1. Effect of word attributes: (a) word length, (b) word familiarity, (c) initial phoneme in words, and (d) similar wards. T. Watanabe, "Quality assessment for synthetic speech using word intelligibility score," *The second joint meeting of ASA and ASJ* (1988).

phoneme is a vowel or a semivowel /w/, /j/ has high word intelligibility, and a word whose initial phoneme is a voiced stop or an affricative /b/,/d/,/g/,/z/,/r/ gives low word intelligibility. The tendency is stronger in TTS speech than natural speech. Another important result is that influence of the initial phoneme differs among the 3 TTS systems, as shown in the fricatives /s/,/h/ and nasals /m/,/n/. As shown in Figure 3.1 (d), the existence of similar words also influences word intelligibility. Here, the similar word was defined as a word belonging to the most familiar word group, and only having one phoneme different from those of the object word.

A learning effect for TTS speech was investigated using word intelligibility tests. The tests were performed during four time periods. One- or two-month intervals were used as the experiment periods. The subjects were four female listeners who had had no previous experience

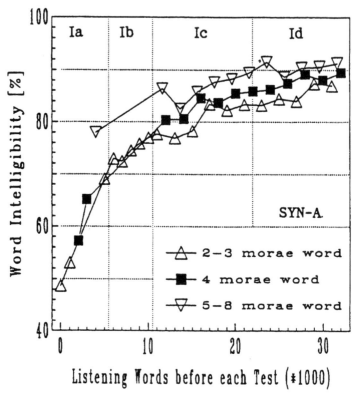

Figure 3.2. Learning effect. T. Watanabe, "Quality assessment for synthetic speech using word intelligibility score," *The second joint meeting of ASA and ASJ* (1988).

in listening to TTS speech. Figure 3.2 shows the word intelligibility score of TTS speech. In the early stage of the tests, the word intelligibility score was only about 50%, but this increased to 70–80% after 10,000 words had been presented. After more than 10,000 words, the score increased more gradually. An interesting point is that subjects' ability to recognize TTS speech did not degrade even if there was a two-month interval between tests.

The difference between trained and inexperienced subjects was investigated using word intelligibility tests. The experiment was

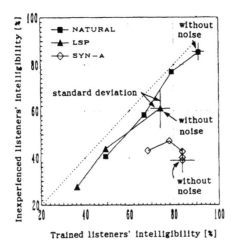

Figure 3.3. Word intelligibility of well-trained listeners vs. that of native listeners. T. Watanabe, "Quality assessment for synthetic speech using word intelligibility score," *The second joint meeting of ASA and ASJ* (1988).

designed to compare the degradation in word intelligibility under several noise conditions using TTS speech, natural speech, and analysis-synthesis speech. In Figure 3.3, the intelligibility scores of the trained and inexperienced subjects are plotted according to noise conditions. For both groups of subjects, the less noise is, the better the intelligibility scores (noise condition is not indicated in the figure). In terms of the intelligibility scores for natural and analysis-synthesis speech, there is a small difference between the two groups, but a large difference is observed in TTS speech. The difference is the learning effect. Therefore, to evaluate TTS speech, an important factor is whether the subjects are trained or inexperienced.

The influence of noise level was investigated for three kinds of TTS using word intelligibility tests. Figure 3.4 shows some of the experimental results. The deterioration rate varies widely among the three kinds of TTS; i.e., the intelligibility of SYN-C is very sensitive for noise level. The result suggests that the influence of noise is an important factor in TTS evaluation.

To evaluate the total performance of TTS, subjective evaluations are necessary, because there is no objective measure or function like signal-

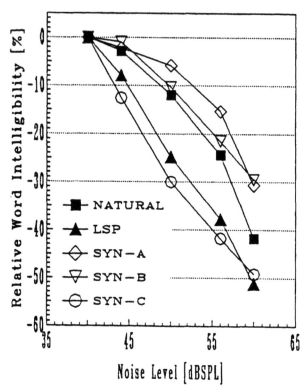

Figure 3.4. Noise level vs. relative word intelligibility under 62 dB SPL without noise. T. Watanabe, "Quality assessment for synthetic speech using word intelligibility score," *The second joint meeting of ASA and ASJ* (1988).

to-noise ratio (SNR) in speech coding. Typical is the diagnostic evaluation method applied to three kinds of TTS [Watanabe-91]. Fifteen items were selected; the synthesized speech [A] is easily understood? [B] sounds to be a different phoneme? [C] has noise? [D] is hard to hear? [E] is naturally concatenated? [F] has natural rhythm? [G] has natural pause? [H] has natural word accent? [I] has natural sentence intonation? [J] is noisy quality? [K] is clear quality? [L] is soft quality? [M] is human-like quality? [N] has natural speaking rate? [O] is good as total impression? To independently evaluate each item, three sentences per

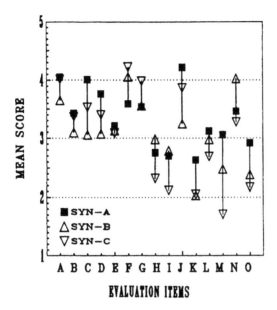

Figure 3.5. Qualities of three kinds of synthetic speech measured by diagnostic assessment method. T. Watanabe, "Global assessment method for synthesized speech (in Japanese)," *Trans. IEICE* Vol. J74-A No. 4 pp. 599–609 (1991).

item were presented to the subjects. They were asked to rate the quality using five categories. Figure 3.5 shows the experimental results. The three TTS have relatively good performance in [A], [D], [F] and [G], but poor performance in [H], [J] and [O]. This and other tests clearly show that there is no TTS that has good performance for all items, so TTS selection depends on usage. The experimental scores were analyzed by factor analysis. The results are shown in Figure 3.6 and Figure 3.7 according to factor importance. Judging from these figures, the most important factor is "ease of understanding" which is related to [A], [B] and [D], the second factor is "quality" which is related to [M] and [L], the third factor is "naturalness in rhythm" which is related to [F] and [G], and the fourth factor is "naturalness in intonation" which is related to [H] and [I]. The experimental results indicate that the 15 factors [A-O] can be roughly reduced into four kinds of factors. This is useful information for designing subjective evaluations.

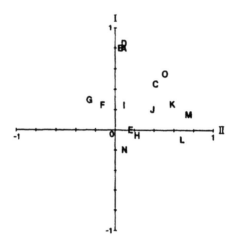

Figure 3.6. Results of factor analysis of diagnostic assessment scores. T. Watanabe, "Global assessment method for synthesized speech (in Japanese)," *Trans. IEICE* Vol. J74-A No. 4 pp. 599–609 (1991).

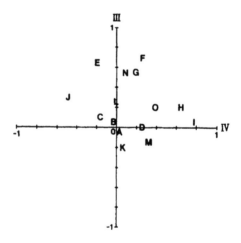

Figure 3.7. Results of factor analysis of diagnostic assessment scores. T. Watanabe, "Global assessment method for synthesized speech (in Japanese)," *Trans. IEICE* Vol. J74-A No. 4 pp. 599–609 (1991).

3.2. A Guideline for TTS Evaluation

Because the importance of the evaluation methodology has been increasing with the importance of TTS technology, a "guideline for speech synthesizer evaluation" was recently proposed by the Speech Input/Output Systems Expert Committee in the Committee on Standardization of Office Automation Equipment of the Japan Electronic Industry Development Association (JEIDA) [Itahashi-95]. The guideline defines the following four evaluations: text analysis evaluation, intelligibility evaluation, naturalness evaluation, and overall quality evaluation. In text analysis evaluation, text-to-phoneme conversion, accent, and pause must be evaluated for word level and sentence levels. Quantitative evaluation measures are also defined. Intelligibility evaluation consists of syllable articulation, word intelligibility, and sentence intelligibility tests. For the syllable articulation test, the following points are defined to specify the evaluation to be performed; utterance units to be evaluated, attributes of evaluators, testing method (such as preliminary training, number of test trials, testing equipment, method of presenting test speech and instructions for writing down), method of summarizing test results (average syllable articulation, confusion matrix and summary for each syllable position) and test conditions to be specified. For the word intelligibility test, the following points are defined to specify evaluation; selection of words (a selection program is available), testing method, evaluation with semantically unpredictable sentences (sentences for evaluation, phrase pattern and so on). For the sentence intelligibility test, texts for evaluation and styles for questions and answers are defined. In terms of naturalness evaluation, the following points are defined to specify the evaluation; evaluation items (such as confusion, clarity, smoothness, noisiness, humanness, rhythm, pause, accent, intonation and overall impression), evaluation category (such as natural, some unnaturalness, a slight feeling of uneasiness, feel uneasy, and extremely uneasy), testing methods, attributes of evaluators, listening conditions and evaluation texts. In terms of overall quality evaluation, the following points are defined to specify evaluation; evaluation items (such as intelligibility, speech sound quality, temporal factors, intonation, overall goodness and suitability), evaluation categories, evaluation method, attributes of evaluators, evaluation text and evaluation conditions.

CHAPTER 4

Towards Flexible Speech Synthesis-By-Rule

As shown in the previous chapter, technologies related to Japanese speech synthesis-by-rule have been studied extensively and significant progress has been made in the last two decades. These efforts have resulted in several text-to-speech software products. Through the distribution of these products, speech synthesis-by-rule will be exposed to serious evaluation by the marketplace. We can assume that users are now demanding synthesized speech that is more natural and human-like. While improving the quality of synthesized speech is quite important, another interesting approach is to provide functions that make it possible to synthesize different speech styles; i.e., to increase the controllability of speech synthesis-by-rule. In this chapter, we will discuss some studies in this direction.

4.1. Speaking Styles [Abe-96]

Research on text-to-speech have concentrated on synthesizing speech in a standard speaking style such as news reading, weather forecasting and so on. This is mainly because the standard speaking style is appropriate for information providing services. However, to extend the application area of synthesized speech, the synthesis of various speech styles is needed. Text-to-speech (TTS) systems will be required to synthesize not only the standard reading style, but also task specific styles such as commercial advertisements and warnings. Moreover, for human/ machine problem-solving dialog systems, the output speech should express information in a more natural manner by reproducing emphasis and emotion. To achieve these goals, one study analyzed three different speaking styles and then synthesized them in a text-to-speech system.

4.1.1. Speech Material

Because, in some cases, speaking style is determined by text content, different texts were used for each speaking style. Three texts from different fields were selected as material to which a distinct style could be clearly assigned; i.e., a paragraph of an artistic novel, advertisement phrases, and a paragraph of an encyclopedia. They were selected not only so that a narrator would consciously say them differently, but also that listeners would easily recognize them as different styles. A set of common text; i.e., a set of 100 sentences and 216 phonetically balanced words were also selected. For speech recording, a professional narrator first spoke the text of a field in an appropriate speaking style that was his own. This constituted a preparation session to fix the speaking style. Just after this session, he uttered the common text in the same speaking style; i.e., 100 sentences and 216 phonetically balanced words. All recorded speech data were manually assigned phonetic transcriptions. To extract the characteristics of speaking styles, we used the common text speech because it was free of text-dependent effects. Hereinafter, for convenience, we refer to the three speaking styles as the novel, advertisement and encyclopedia speaking styles.

4.1.2. Spectral Characteristics of Different Speaking Styles

To investigate the spectral characteristics of the speaking styles, we applied a voice conversion algorithm based on codebook mapping (see 4.2.1). Six mapping codebook pairs (all combinations of the 3 styles) were generated. The spectral parameters of formant frequencies (F_1, F_2, F_3), formant bandwidth, and spectrum tilt were extracted from each codevector. Formant frequencies were determined by referring to the pole frequencies of an AR model, and formant correspondences between codevectors were manually determined. Spectrum tilt was defined as the first-order regression coefficients of the LPC spectrum envelope. Phonetic attributes were also assigned to codevectors using phonetic transcriptions.

Characteristics of Formant Frequency

Figure 4.1 shows the F_1–F_2 plane of the three speaking styles. In all vowels, the 1st formant frequency increases in the order of novel, encyclopedia, and advertisement speaking style. The difference in F_1 frequency is about 10% to 20% of the F_1 value of the encyclopedia

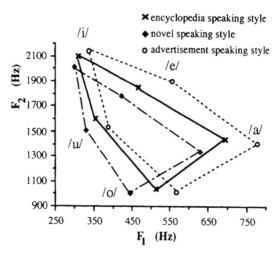

Figure 4.1. F1–F2 plane of the three speaking styles. M. Abe, "Speaking styles: statistical analysis and synthesis by a text-to-speech system," *Progress in speech synthesis*, Springer, pp. 495–510 (1996).

speaking style. On the other hand, the F_2 differences among the three styles are about 5%. In terms of the 3rd formant frequency, data is not shown here, the novel speaking style has the lowest frequency, and the difference is 5–10%. A previous study showed that voice-personality is perceptually lost when an individual formant is shifted 15% towards both the high and the low frequency region. Judging from the reference, in terms of F_1 and F_3, formant differences among speaking style is so large as to lose voice-personality. These results suggest that modifying formant frequency is important to synthesize realistic speaking styles.

Characteristics of Spectral Tilt

The same trends in spectral tilt characteristics were observed in all vowels. For example, Figure 4.2 shows the tendency of the phoneme / o/. Each circle represents the spectral tilts extracted from a pair of associated codevectors. The empty circles plot the relationship between the encyclopedia speaking style and the novel speaking style, and the filled circles show the relationship between the encyclopedia speaking style and the advertisement speaking style. Regression lines are drawn

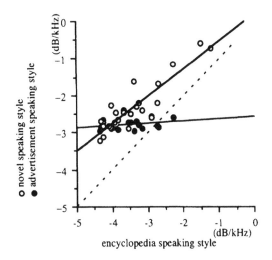

Figure 4.2. Spectral tilts in phoneme /o/. M. Abe, "Speaking styles: statistical analysis and synthesis by a text-to-speech system," *Progress in speech synthesis*, Springer, pp. 495–510 (1996).

for both empty and filled circles. The spectral tilt of the novel speaking style is always flatter than that of the encyclopedia speaking style. The spectral tilt values of the novel and encyclopedia speaking styles are distributed from −3 dB/kHz to −1 dB/kHz, and the distribution is considered to be caused by slight spectrum differences in /o/. This may be because speech is produced in a free and easy manner in the novel and encyclopedia speaking styles. On the other hand, the spectral tilts of the advertisement style are almost constant at around −3 dB/kHz. This suggests that articulation is less variable because advertisement speaking style is done louder, with bigger jaw opening. This could be explained by hyperarticulation.

4.1.3. Prosodic Characteristics of Different Speaking Styles

Characteristics of Fundamental Frequency

To investigate fundamental frequency (F_0) characteristics of the three speaking styles, a global-local model was applied (refer pp.19). The global models were generated for each speaking style. Table 4.1

Table 4.1. Partial and multiple correlation coefficients of a global model for the three speaking styles. M. Abe, "Speaking styles: statistical analysis and synthesis by a text-to-speech system," *Progress in speech synthesis*, Springer, pp. 495—510 (1996).

Speaking Styles	Partial Correlation Coefficients							Multiple Correlation Coefficient	RMS Error (Hz)	Average F_0 (Hz)
	Boundary		Accent Type of the Phrase			Syllable Number in Current Phrase	Part of Speech			
	Preceding	Following	Preceding	Current	Following					
encyclopedia	0.417	0.528	0.130	0.337	0.018	0.385	0.231	0.834	18.79	161.7
advertisement	0.475	0.537	0.110	0.221	0.039	0.323	0.150	0.804	25.55	229.6
novel	0.311	0.438	0.226	0.167	0.094	0.246	0.159	0.689	16.86	121.7

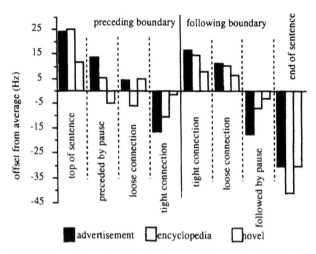

Figure 4.3. Effects of boundary for the three speaking styles. M. Abe, "Speaking styles: statistical analysis and synthesis by a text-to-speech system," *Progress in speech synthesis*, Springer, pp. 495–510 (1996).

shows the parameters of the global model for the three speaking styles. The average F_0 indicates that the F_0 ranges are quite different for the three speaking styles. The multiple correlation coefficient means the model's preciseness. Judging from the values of encyclopedia and advertisement speaking styles (0.834 and 0.804, respectively), good global models were constructed for both styles. On the other hand, the global model for the novel speaking style is relatively poor (0.689). This result implies that other factors such as emotion, speaker's intention and so on are also important in forming the novel speaking style. The partial correlation coefficient indicates how important a factor is in a model. The experimental results show that the importance of the factors are the same in all speaking styles; i.e., boundary type is most important, syllable number is 2nd most, and so on. However, the components of the boundary type factor have quite different values for each speaking style. Figure 4.3 shows the values of boundary factors. Judging from Figure 4.3, the main differences are that the dynamic range of the novel speaking style is narrower than those of the others, and that the phrase height of the advertisement speaking style differs from those of the others when the phrase is preceded or followed by pause.

The local model is mainly controlled by accent types and phonemes. These factors might be only slightly changed in different speaking styles. For example, in Japanese, changing an accent type changes the meaning of a word. Therefore, we have a hypothesis that a single local model is applicable for various speaking styles. To confirm this point, the local model was modified to output the difference F_0 value between the encyclopedia speaking style and the other speaking styles. The encyclopedia style was used as a reference. Because we have engaged in synthesizing the encyclopedia style in the conventional TTS system, this comparison is useful. Table 4.2 shows the local model parameters. In all models, the multiple correlation coefficients are quite high, and in terms of the partial correlation coefficient, the phrase height difference factor is dominant. This indicates that accent type and phoneme have similar influence in the three speaking styles. The results confirm the hypothesis that a single local model is applicable for all three speaking styles, only the phrase height differences for each speaking style need be controlled.

Judging from these results, we can conclude that one local model can be applied to the three speaking styles, but a unique global model should be constructed for each speaking style.

Characteristics of Segmental Duration

General Characteristics

To investigate the general characteristics of segmental duration, sentence duration was compared for the three speaking styles. The analysis considered only those sentences within which pauses were inserted at the same position in each speaking style (60 sentences). Figure 4.4 shows the ratio of sentence duration in the novel or advertisement speaking style to the sentence duration in the encyclopedia speaking style.

When sentence duration is calculated after eliminating pauses, the ratio shows the average lengthening or shortening of phoneme duration. Compared to the encyclopedia speaking style, on average, the novel speaking style has longer phoneme duration (1.1 times), while that of the advertisement style is shorter (0.9 times). The existence or absence of pause duration causes more difference in the novel speaking style than in the advertisement style. That is, the novel speaking style has much longer sentence duration (1.2 times) when the sentence includes pauses.

Table 4.2 Partial and multiple correlation coefficients of a local model. M. Abe, "Speaking styles: statistical analysis and synthesis by a text-to-speech system," *Progress in speech synthesis*, Springer, pp. 495–510 (1996).

| SBU models | Tone Pattern | Partial Correlation Coefficients | | Max Difference F_0 in a phase | Position Within a Phase | Multiple Correlation Coefficient | RMS error (Hz) (Hz/3ms) |
| | | Consonant Class | | | | | |
		Current Syllable	Following Syllable				
1st syllable	0.25	0.13	0.09	0.97	—	**0.87**	18.4
2nd syllable	0.09	0.06	0.07	0.97	—	**0.97**	13.3
3rd syllable	0.21	0.12	0.10	0.98	—	**0.97**	13.7
4th syllable	0.15	0.06	0.17	0.95	—	**0.95**	16.7
other syllables	0.14	0.21	0.14	0.93	0.14	**0.93**	18.1

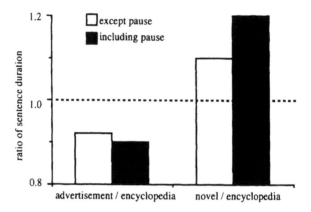

Figure 4.4. Sentence duration for the three speaking styles. M. Abe, "Speaking styles: statistical analysis and synthesis by a text-to-speech system," *Progress in speech synthesis*, Springer, pp. 495–510 (1996).

This indicates that speaker emphasized the pause length to decrease the speech rate.

Vowel duration with different syllable positions

For the 60 sentences analyzed in the previous section, vowel duration was classified for several different syllable positions such as the beginning or end of the sentence, preceded or followed by pause and others. Figure 4.5 shows the average duration and standard deviation for the three speaking styles. When a syllable is followed by pause, vowel duration is lengthened. The increase is about 80 msec in both encyclopedia and advertisement speaking styles, but the increase is much longer (about 150msec) in the novel speaking style. Moreover, in that style, vowel duration is lengthened (about 40msec) when the syllable is located at the sentence end. Judging from the above results, we can conclude that syllable lengthening ahead of a pause is an important phenomena in recreating the novel speaking style.

Characteristics of Speech Power

It has been proposed to control speech power according to the fundamental frequency. Speech power was examined in the same way.

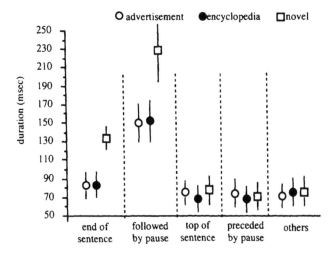

Figure 4.5. Average duration and standard deviation of vowels in different syllable position. M. Abe, "Speaking styles: statistical analysis and synthesis by a text-to-speech system," *Progress in speech synthesis*, Springer, pp. 495–510 (1996).

Figure 4.6 shows the relationship between the average power of a vowel segment and F_0 value in the center of the vowel. In general, the higher the F_0, the larger the speech power. The relationship is almost linear in the range from 100 Hz to 200 Hz, but the speech power increase saturates above 200 Hz. It is also interesting that the speaker used the range above 200 Hz only in the advertisement speaking style. This confirms that the advertisement speaking style was uttered in an extreme way. Judging from the results, it is reasonable to control speech power according to F_0.

4.1.4. A Strategy for Changing Speaking Styles in TTS

According to the analysis results given in the previous sections, a strategy was proposed to separate rules into two groups: specific rules (Srules), specific to a particular speaking style, and general rules (Grules) which are applied to all speaking styles. This strategy makes it easy to adapt a TTS system to any particular speaking style, because only Srules need be changed to match the speaking style required. Grules

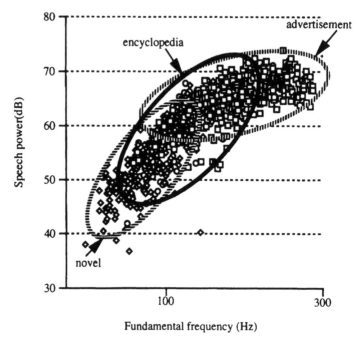

Figure 4.6. Relationship between F_0 and power. M. Abe, "Speaking styles: statistical analysis and synthesis by a text-to-speech system," *Progress in speech synthesis*, Springer, pp. 495–510 (1996).

are the rules of encyclopedia speaking style, because this speaking style is often used to develop conventional TTS systems. Srules are as follows:

In terms of formant frequencies, (1) the value of the 1st formant frequency is decreased 10% and increased 20% for the novel speaking style and the advertisement speaking style, respectively, and (2) the value of the 3rd formant frequency is decreased 20% for the novel style. In terms of fundamental frequency (F_0), the global model is generated as an Srule for each style, but no Srule is necessary for the local model. In terms of phoneme duration, (1) average phoneme duration of the novel speaking style is longer (1.1 times), while that of the advertisement style is shorter (0.9 times), (2) in the novel speaking style, duration lengthening for syllables followed by a pause is emphasized 1.8 times, (3) in the novel speaking style, vowel duration is lengthened (about 40 msec) when the syllable is located at the sentence end, and (4) in the novel speaking style, pause duration is lengthened. Speech power is controlled according to F_0. The ratio is approximated using the output from the F_0 global model.

Evaluation by Listening Test

An ABX listening test was carried out to check how effective the rules were. In addition to synthesizing speech in different speaking styles, using the same rules human speech uttered in the encyclopedia speaking style was converted to other speaking styles. Stimuli A and B were speech uttered in the encyclopedia and the target speaking style respectively, and stimulus X was synthesized either in the target speaking style or speech converted to the target speaking style from the encyclopedia speaking style or the speech uttered in the encyclopedia or the target speaking style. Different sentences were used for A, B, and X. Two sets of three different sentences were used for human speech conversion and speech synthesis, and the total number of stimuli was 36. Eight listeners were asked to select either A or B as being closest to X.

Figure 4.7 shows the experimental results; Figure 4.7(a) is for human speech conversion, Figure 4.7(b) is for synthetic speech. Even in the case of human speech, speaking styles are not always judged to match the speaker's intention (93.8%, 95.8% and 83.3%). This indicates the difficulty of this task in some sense. Referring to the performance in human speech identification, the synthesized speech and converted human speech were effectively reproduced as the target speaking styles. These results indicate that the Srules are effective in generating several speaking styles.

4.2. Voice Quality Control

With regard to flexible speech synthesis-by-rule, controllability of voice quality is another important aspect. In fact, users demand that a TTS be able to synthesize speech in different voice qualities. For example, it was reported that when a TTS was used for long periods, users wanted the voice quality to change from time to time. When the same TTS is used in several commercial systems such as automatic account inquiries, ticket vending machine and so on, the companies often want the voice

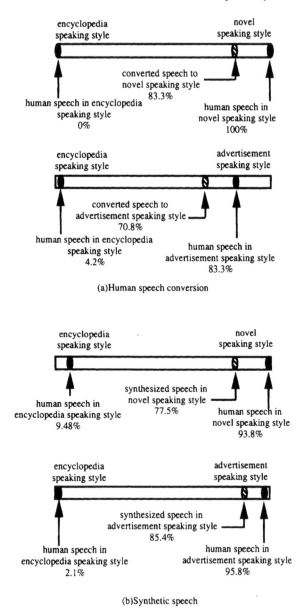

(a)Human speech conversion

(b)Synthetic speech

Figure 4.7. ABX experiment results. M. Abe, "Speaking styles: statistical analysis and synthesis by a text-to-speech system," *Progress in speech synthesis*, Springer, pp. 495–510 (1996).

quality to match the corporate identity. In a station announcement system, assigning different voice quality to different lines makes it possible to avoid confusion when several announcements occur at the same time.

However, the voice quality controllability of TTS commercial products are limited and most allow us only to select either a male-like voice or female-like voice. This is mainly because of the recent approach; i.e., to synthesize speech by concatenating speech synthesis units after modifying its prosodic parameters based on time domain algorithm. As explained in Chapter 2, time domain algorithms make it possible to synthesize high quality speech, but they offer no way of changing the spectrum parameters that must be modified for voice quality control. One possible solution is to collect speech data uttered by many speakers and to construct a set of synthesis units for each speaker. This is time consuming and requires a large storage for synthesis units.

To solve this problem, "voice conversion" is an interesting approach. Voice conversion is achieved by an algorithm that changes or modifies speaker individuality; i.e., speech uttered by one speaker is transformed in order to sound as if it had been articulated by another speaker. By applying this approach to synthesis units, TTS makes it possible to change voice quality. Besides this, the approach lends itself to a variety of significant applications: designing hearing aids appropriate for specific hearing problems, improving the intelligibility of abnormal speech uttered by a speaker who has speech organ problems, etc.

4.2.1. Voice Conversion Based on Codebook Mapping [Abe-90]

A concept was proposed for treating voice conversion as a mapping problem in the spectrum space. This idea was originally proposed to adapt speech recognition models to an unknown speaker. Figure 4.8 shows the basic idea of voice conversion using vector quantization. Ellipses in Figure 4.8 represent the codebooks (spectrum spaces) of speaker A and speaker B, and the black dots in each ellipse are code vectors (speech spectra). Let's try to convert speaker A's speech to speaker B's speech. If the code vectors in these codebooks have one-to-one correspondences, like code vector A_1 and code vector B_1 in Figure 4.8, voice conversion is easily performed by replacing A_1 with B_1. In other words, the conversion of acoustic features from one speaker to another is reduced to the problem of finding the

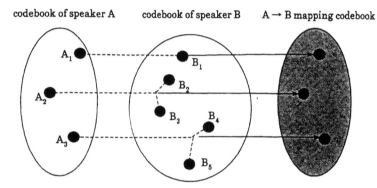

Figure 4.8. Basic idea of voice conversion.

correspondence between the codebooks of the two speakers. However, as with A_2-A_3 and B_2-B_5, code vectors usually do not have one-to-one correspondences. Therefore, a key point is to generate "mapping codebooks" that is a pair of codebooks, where the code vectors of one codebook (original speaker's) have a one-to-one correspondence with the code vectors of the other codebook (target speaker's). In an off-line procedure, the mapping codebooks are generated once in advance between two speakers, so the on-line procedures can realize voice conversion.

Off-line Procedures

Figure 4.9 shows a block diagram of the procedure for generating mapping codebooks for spectrum parameters. Mapping codebooks are generated as follows:

Step 1:   Using training utterances, generate a speaker-specific codebook for each speaker, say Speaker A and Speaker B, using the LBG algorithm. Here, both speakers must say the training utterances. Speaker A's and Speaker B's utterances are then vector quantized using his/her codebook.

Step 2:   Using the code vector sequence for the same text utterance from the two speakers, the correspondence between the code vectors are determined by Dynamic Time Warping.

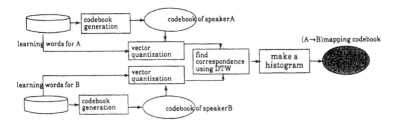

Figure 4.9. Method for generating a mapping codebook. M. Abe, S. Nakamura, H. Kuwabara, K. Shikano, "Voice conversion through vector quantization," *Jour. of Acoust. Soc. of Japan*, Vol. E-11, No. 2, pp. 71–76 (1990).

Step 3:   The code vector correspondences between the two speakers are stored as histograms for all training utterances.

Step 4:   Using the histogram for each code vector of Speaker A as the weighting function, a mapping codebook from Speaker A to B is defined as the linear combination of Speaker B's vectors.

Step 5:   Steps 2, 3, and 4 are repeated to refine the mapping codebook.

Pitch frequencies and power values contribute heavily to speech individuality. Mapping codebooks for these parameters are also generated at the same time using the same procedure mentioned above except (1) pitch frequencies and power values are scalar-quantized, and (2) the mapping codebook for pitch frequencies is defined based on the maximum occurrence in the histogram.

On-line Procedures

Figure 4.10 shows a block diagram of the on-line procedures.

Step 1:   Speaker A's speech is analyzed by the linear prediction method.

Step 2:   Input spectrum parameters X are fuzzy vector-quantized as X' using his/her codebook. The fuzzy VQ approximates an input vector as the linear combination of code vectors and can represent input vectors more precisely than conventional VQ because it escapes the limitation caused by codebook size. Fuzzy VQ is defined as follows:

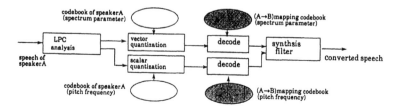

Figure 4.10. Block diagram for voice conversion from speaker A to speaker B.
M. Abe, S. Nakamura, H. Kuwabara, K. Shikano, "Voice conversion through
vector quantization," *Jour. of Acoust. Soc. of Japan*, Vol. E-11, No. 2, pp. 71–
76 (1990).

$$X' = \frac{\sum_{l=1}^{k}[(u_i)^m \times V_i]}{\sum_{l=1}^{k}(u_i)^m}$$

$$= \sum_{l=1}^{k}\left[\frac{(u_i)^m}{\sum_{l=1}^{k}(u_i)^m}\right] \times V_i$$

$$= \sum_{l=1}^{k}u_i' \times V_i$$

u_i is fuzzy membership function. $u_i \in [0,1]^{\forall} i$.

$$u_i = \frac{1}{\sum_{j=1}^{k}\left(\frac{\|X - V_i\|}{\|X - V_j\|}\right)^{\frac{1}{m-1}}}$$

V_i is a code vector in speaker A's codebook V.
m : fuzziness.
k : number of the nearest code vectors.

Parameters for pitch frequencies and power values are scalar-quantized using his/her codebooks.

Step 3:   Spectrum parameter conversion is performed by replacing
          speaker A's code vectors by speaker B's code vectors using
          mapping codebooks as follows:

$$X^{map} = \sum_{i=1}^{k} u_i' \times V_i^{map}$$

where V^{map} is speaker B's code vector in the mapping codebook
and X^{map} is a converted vector. This formulation assumes that
fuzzy membership function $u\phi_i$ determined in speaker A's
spectrum space is preserved in speaker B's spectrum space.
This is reasonable because the spectrum space is considered
to be smooth in local spectrum space. Parameters for pitch
frequencies and power values are decoded using the mapping
codebooks.

Step 4:   Speech is synthesized by an LPC vocoder. The output speech
          will have the voice individuality of speaker B.

4.2.2. Voice Conversion Based on Piecewise Linear Conversion Rules of Formant Frequency [Mizuno-95]

The use of vector quantization was found to be effective in generating
conversion rules that were statistically guaranteed to be reasonable.
However, this degraded the quality of the converted speech because the
variability of the output signal pattern was restricted by the size of the
codebook. To solve this problem, an algorithm was proposed to cope
with mapping codebooks. Key points of the algorithm are (1) formant
frequency and spectrum tilt are used to change speaker individuality, (2)
rules for changing formant frequencies are generated by referring to
mapping codebooks, (3) voice conversion is performed pitch-
synchronously, and (4) the residual signal is used as voice source. The
size of the mapping codebooks does not restrict the range of the output
signal pattern, only the number of formant conversion rules.
Consequently, in terms of total voice conversion, the new voice
conversion algorithm synthesizes speech of higher quality than the
conventional algorithm.

Making Formant Frequency Conversion Rules (off-line procedures)

Formant frequency conversion rules are generated using the mapping
codebooks. Figure 4.11 shows the LPC spectrum envelopes that were

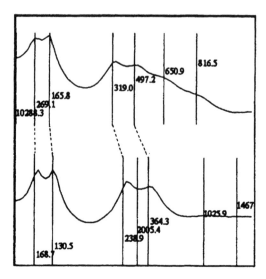

Figure 4.11. The correspondence of the poles extracted from the mapping codebooks. H. Mizuno, M. Abe, "Voice conversion algorithm based on piecewise linear conversion rules of formant frequency and spectrum tilt," *Speech Communication* **16**, Elsevier Science – NL, pp. 153–164 (1995).

extracted from the corresponding code vectors of a mapping codebook. The upper one is the target speaker's spectrum envelope and the lower one is that of the original speaker. In Figure 4.11, pole frequencies are indicated by vertical lines, and the numbers indicate the bandwidth of each pole. Taking into account the pole bandwidth by a direct visual inspection, certain poles are selected as formants (F_1, F_2, F_3, and F_4) for both the target and the original speaker. The formant correspondences are manually determined (the resulting correspondence is shown by the slanted lines in Figure 4.11). At this stage, in order to correctly assign formant correspondence, this process is carried out manually. Spectrum tilt is also obtained from the corresponding code vector. The original speaker's code vector, the formant frequency shift values and spectrum tilt shift values are stored as the conversion rules. The extracted formant frequency is also stored as the reference formant frequency. Figure 4.12 shows the formant frequency shift rules on the F_1–F_2 plane for two male speakers. The arrows indicate formant shifts from the original speaker to the target speaker. Formant frequency shift rules are precisely

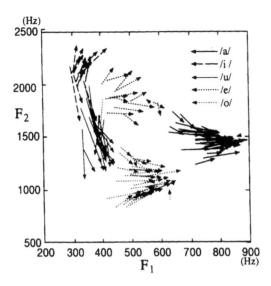

Figure 4.12. Conversion rules of F_1 and F_2. H. Mizuno, M. Abe, "Voice conversion algorithm based on piecewise linear conversion rules of formant frequency and spectrum tilt," *Speech Communication* **16**, Elsevier Science – NL, pp. 153–164 (1995).

described for each vowel, and the streams of the shift directions indicate quite reasonable conversion.

Voice Conversion Algorithm (on-line procedures)

Figure 4.13 shows a block diagram of the proposed algorithm. In the following explanation, the numbers refer to the block numbers cited in Figure 4.13.

[1] Input speech uttered by the original speaker is vector quantized using the original speaker's codebook.

[2] Outputs of [1] are smoothed by using the code vector buffer. This buffer includes the current code vector and the three previous code vectors. The code vector which occurs most frequently in the buffer is selected as an output code vector. This smoothing technique results in the accurate extraction of formant frequencies in [3] and the accurate generation of target formant frequencies in [4].

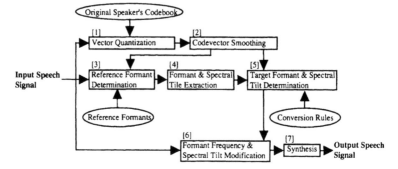

Figure 4.13. Block diagram of voice conversion algorithm. H. Mizuno, M. Abe, "Voice conversion algorithm based on piecewise linear conversion rules of formant frequency and spectrum tilt," *Speech Communication* **16**, Elsevier Science – NL, pp. 153–164 (1995).

[3] Depending upon the output code vector, reference formant frequencies are set using the stored data generated in off-line procedures.

[4] LPC poles and spectrum tilts are calculated from the input speech. If a pole has a narrow bandwidth, and the pole frequency is the nearest to the reference formant frequency, the LPC poles are selected as formant.

[5] The target formant frequency is the sum of the extracted formant frequency and formant shift value which is determined by the associated code vector. The target spectral tilt is obtained in the same way. The target spectral intensity is the value of the input spectral intensity.

[6] The formant frequencies and spectrum tilt are iteratively modified in order to minimize the spectral distance against the target values.

[7] The speech signal is synthesized by pitch-synchronous overlap addition.

4.2.3. Performance Evaluation

Four speakers provided speech material; three were men (man1, man2 and man3) and one was an 11-year-old boy. In all tests, man1's voice was converted to match those of the others. Mapping codebooks were generated for these speaker pairs. Table 4.3 illustrates the spectrum

Table 4.3. The average F_0 difference and cepstrum distortion at mapping codebook generation. H. Mizuno, M. Abe, "Voice conversion algorithm based on piecewise linear conversion rules of formant frequency and spectrum tilt," *Speech Communication* 16, Elsevier Science — NL, pp. 153—164 (1995).

Target Speaker	Difference of F_0 (Hz)	CD
boy	93.3	0.84
man2	40.4	0.61
man3	23.5	0.56

distortion and average F_0 differences between man1 and the others. In terms of speech quality, man3 was closest to man1, while the 11-year-old boy was farthest.

A preference test was designed in order to evaluate the speech quality of both the mapping codebooks and the piecewise linear conversion rules. Each test pair consisted of the converted speech produced by the two algorithms. Three different words were used in the conversions. Listeners were asked to indicate their preference for each pair. Figure 4.14 shows the experimental results. These results confirm that the piecewise linear conversion rules have significantly better performance than mapping codebooks. This is mainly due to pitch-synchronous processing and the usage of the residual signal in the piecewise linear conversion rules.

In order to evaluate the overall performance, an opinion test was carried out. Each speech pair consisted of two different words from 4 different groups. The groups were man1's speech, man2's speech, the speech converted by the two algorithms (man1's to man2's). Listeners were asked to rate the similarity of each pair in five categories, ranging from "similar" to "dissimilar". Hayashi's fourth method of quantification was applied to the experimental data obtained by the listening test. This method places a sample in a space according to the similarities between the samples. The representation of the results on a two-dimensional illustration is shown in Figure 4.15. This figure shows the relative similarity-distance between stimuli. Contribution factors of the first axis and second axis were 0.74 and 0.26, respectively. The speech converted by the piecewise linear conversion rules is much closer to the target than the speech converted by the mapping codebooks. One of the reasons is that, as shown in preference test, the piecewise linear conversion rule algorithm synthesizes high quality speech.

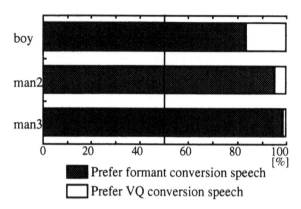

Figure 4.14. Preference score. H. Mizuno, M. Abe, "Voice conversion algorithm based on piecewise linear conversion rules of formant frequency and spectrum tilt," *Speech Communication* **16**, Elsevier Science – NL, pp. 153–164 (1995).

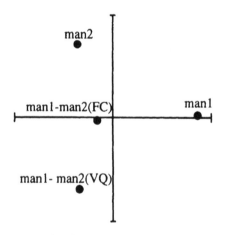

man1: original man1's speech
man2: original man2's speech
man1-man2(VQ): VQ conversion
man1-man2(FC)   :formant conversion

Figure 4.15. Distribution of psychological distance for man1-to-man2 voice conversion. H. Mizuno, M. Abe, "Voice conversion algorithm based on piecewise linear conversion rules of formant frequency and spectrum tilt," *Speech Communication* **16**, Elsevier Science – NL, pp. 153–164 (1995)

References

[Abe-90] M. Abe, S. Nakamura, H. Kuwabara, K. Shikano, "Voice conversion through vector quantization," *Jour. of Acoust. Soc. of Japan*, Vol. E-**11**, No. 2, pp. 71–76 (© 1990 IEEE).

[Abe-92] M. Abe, H. Sato, "Two-stage F_0 control model using syllable based F_0 unit," *Proc. ICASSP92*, Vol. 2 pp. 53–56 (1992).

[Abe-96] M. Abe, "Speaking styles: statistical analysis and synthesis by a text-to-speech system," *Progress in speech synthesis*, Springer, pp. 495–510 (1996).

[Allen-87] J. Allen, M. Hunnicutt, D. Klatt, *From text to speech; The MITalk system*, Cambridge Univ. Press (1987).

[Arai-94] Y. Arai, T. Minowa, H. Yoshida, H. Nishimura, H. Kamata, T. Honda," A study on the high quality voice editing system (in Japanese)," *Technical report of IEICE*, SP93–141, pp. 23–30 (1994).

[Fujisaki-71] H. Fujisaki, H. Sudo, "Synthesis by rule of prosodic features of connected Japanese," *Proc. of 7th ICA*, Vol. **3**, pp. 133–136 (1971).

[Fujisaki-84] H. Fujisaki, K. Hirose, N. Takahashi, M. Yokoo, "Realization of accent components in connected speech (in Japanese)," *Trans. of the committee on speech research, Acoust. Soc, Jpn.*, S84–36 (1984).

[Hakoda-76] K. Hakoda, H. Sato, "Some characteristics of prosodic pattern in spoken sentences (in Japanese)," *Trans. of the committee on speech research, Acoust. Soc, Jpn.*, S76–19 (1976).

[Hakoda-80] K. Hakoda, H. Sato, "Prosodic rules in connected speech synthesis," *Systems, Computers, Controls*, Vol. **11**, No. 5, pp. 28–37 (1980).

[Hakoda-95] K. Hakoda, T. Hirokawa, H. Tsukada, Y. Yoshida, H. Mizuno, "Japanese text-to-speech software based on waveform concatenation method," *Proc.of AVIOS'95*, pp. 65–72 (1995).

[Hashimoto-74] S. Hashimoto, S. Miyahara, "A method of linear approximation model of pitch contour (in Japanese)," *Trans. of the committee on speech research, Acoust. Soc, Jpn.*, S74–11 (1974).

[Hayashi-50] C. Hayashi, "On the quantification of qualitative data from the mathematicostatistical point of view," *Ann. Inst. Statist. Math* **2** (1950).

[Higuchi-80] N. Higuchi, H. Fujisaki, "Duration control of segmental features in connected speech (in Japanese)," *Trans. of the committee on speech research, Acoust. Soc, Jpn.*, S80–40 (1980).

[Higuchi-81] N. Higuchi, H. Fujisaki, "Influence of neighboring phonemes upon duration of vowels in connected speech (in Japanese)," *Trans. of the committee on speech research, Acoust. Soc, Jpn.*, S80–96 (1981).

[Hirai-94] T. Hirai, N. Iwahashi, N. Higuchi, Y. Sagisaka, "Auto classification of F0 control commands using statistical analysis," *Technical report of IEICE*, SP94-12, pp. 47–54 (1994).

[Hirose-79] K. Hirose, H. Fujisaki, K. Ohta, "Acoustic features of the fundamental frequency contours of declarative sentence in Japanese (in Japanese)," *Trans. of the committee on speech research, Acoust. Soc, Jpn.*, S79–03 (1979).

[Hirose-81] K. Hirose, N. Takahashi, H. Fujisaki, "Analysis, synthesis, and perception of fundamental frequency contours in spoken sentences (in Japanese)," *Trans. of the committee on speech research, Acoust. Soc, Jpn.*, S81–36 (1981).

[Hirose-84] K. Hirose, H. Fujisaki, M. Yamaguchi, M. Yokoo, "Synthesis of fundamental frequency contours of Japanese sentences based on syntactic structure (in Japanese)," *Trans. of the committee on speech research, Acoust. Soc, Jpn.*, S83–70 (1984).

[Indurkhya-91] N. Indurkhya, S. Weiss, "Interative rule induction methods," *Applied Intelligence*, Vol. 1, pp. 43–54 (1991).

[Itahashi-95] S. Itahashi, "Guideline for speech synthesizer performance evaluation (in Japanese)," *Acoust. Soc. Jpn. meeting, 3-4-11*, pp. 295–296 (1985).

[Itoh-93] K. Itoh, T. Hirokawa, H. Sato, "Phoneme power control for speech synthesis," *IEICE Trans. Fundamentals*, Vol. E76-A, No. 11, pp. 1911–1917 (1993).

[Iwahashi-93a] N. Iwahashi, Y. Sagisaka, "Duration modelling with multiple split regression," *Proc. Eurospeech93*, pp, 329–332 (1993).

[Iwahashi-93b] N. Iwahashi, N. Kaiki, Y. Sagisaka," Speech segment selection for concatenative synthesis based on spectral distortion minimization," *IEICE Trans. Fundamentals*, Vol. E76-A, No. 11, pp. 1942–1948 (1993).

[Kaiki-92] N. Kaiki, K. Takeda, Y. Sagisaka, "Linguistic properties in the control of segmental duration for speech synthesis," *Talking Machines: Theories, Models, and Designs*, G. Bailly, C. Benoit, and T. R. Sawalis (Editors), Elsevier Science Publishers, pp. 180–188 (1992).

[Katae-95] N. Katae, T. Matsumoto, S. Kimura, M. Kaseda, T. Ohyama, "High-quality Japanese text-to-speecch system: NARSYS," *Proc. Eurospeech95*, pp. 1861–1864 (1995).

[Kawai-93] H. Kawai, N. Higuchi, T. Shimizu, S. Yamamoto, "A study of a text-to-speech system based on waveform splicing (in Japanese)," *Technical report of IEICE*, SP93–9, pp. 49–54 (1993).

[Kindaichi-81] H. Kindaichi, K. Akinaga, *Japanese accent dictionary*, Tokyo: Sanseido, (1981).

[Markel-76] J. Markel, A. Gray, *Linear Prediction of Speech*, Springer-Verlag (1976).

[Mimura-91] K. Mimura, N. Kaiki, Y. Sagisaka, "Statistically derived rules for amplitude and duration control in Japanese speech synthesis," *Proc. Korea-Japan joint symposium on acoustics*, pp. 151–156 (1991).

[Mitome-94] Y. Mitome, "Speech analysis technique for PSOLA synthesis based on complex cepstrum analysis and residual excitation," *Proc. ICSLP94*, pp. 1075–1078 (1994).

[Mizuno-95] H. Mizuno, M. Abe, "Voice conversion algorithm based on piecewise linear conversion rules of formant frequency and spectrum tilt," *Speech Communication* 16, Elsevier Science–NL, pp. 153–164 (1995).

[Mochida-93] T. Mochida, T. Kobayashi, K. Shirai, "Speech synthesis of Japanese sentences using large waveform data-base," *Technical report of IEICE*, SP93-91, pp. 13–18 (1993).

[Moulines-90] E. Moulines, F. Charpentier, "Pitch-synchronous waveform processing techniques for text-to-speech synthesis using diphones," *Speech communication* 9, pp. 453–467 (1990).

[Nakajima-88] S. Nakajima, H. Hamada, "Automatic generation of synthesis units based on context oriented clustering," *Proc. ICASSP88*, pp, 659–662 (© 1988 IEEE).

[Quinlan-86] J. R. Quinlan, "Induction of decision trees," *Machine Learning*, 1, pp. 81–106 (1986).

[Sagayama-83] S. Sagayama, K. Kogure, "Japanese text analysis for speech synthesis (in Japanese)," *Trans. of the committee on speech research, The acoustical society of Japan*, S82-78, pp. 617–624 (1993).

[Sagisaka-80] Y. Sagisaka, Y. Tohkura, "Characteristics of segmental durations in connected speech (in Japanese)," *Trans. of the committee on speech research, Acoust. Soc, Jpn.*, S80-34, pp. 267–273 (1980).

[Sagisaka-81] Y. Sagisaka, Y. Tohkura, "Rule of segmental durations using statistical features of segment (in Japanese)," *Trans. of the committee on speech research, Acoust. Soc, Jpn.*, S80-72, pp. 561–568 (1981).

[Sagisaka-82] Y. Sagisaka, H. Sato," Speech synthesis rules in Japanese text to speech system (in Japanese)," *Acoust. Soc. Jpn. meeting*, 1-5-19 (1982).

[Sagisaka-83] Y. Sagisaka, H. Sato," Accentuation rules for Japanese word concatenation (in Japanese)," *Trans. IECE*, 66D, pp. 849– (1983).

[Sagisaka-84] Y. Sagisaka, H. Sato," Accentuation rules for Japanese text to speech conversion (in Japanese)," *Rev. ECL*, 32, pp. 188– (1984).

[Sagisaka-86] Y. Sagisaka, "A study of prosodic parameter control for speech synthesis," Phd. thesis (1986).

[Sakamoto-95] M. Sakamoto, T. Saito, K. Suzuki, Y. Hashimoto, M, Kobayashi, "A new waveform overlap-add technique for text-to-speech synthesis (in Japanese)," *Technical report of IEICE*, SP95-6, pp. 39–45 (1995).

[Sato-77] H. Sato, "Segmental duration and timing location in speech (in Japanese)," *Trans. of the committee on speech research, Acoust. Soc, Jpn.*, S77-31 (1977).

[Saito-92] S. Saito (ed.) *Speech science and technology*, Tokyo Ohm Co. (1992).

[Sato-77] H. Sato, K, Hakoda, "Speech synthesis based on stored speech segments and rules (in Japanese)," *ECL Tech. Jour.*, 27, 12. pp. 2551 (1977).

[Sato-78] H. Sato, "Speech synthesis on the basis of PARCOR-VCV concatenation units (in Japanese)," *Trans. IEICE* '78/11 Vol. 61-D No. 11, pp. 858–865 (1978).

[Sato-84] H. Sato, "Speech synthesis using CVC concatenation units and excitation waveform elements (in Japanese)," *Trans. of the committee on speech research, Acoust. Soc, Jpn.*, S83-69, pp. 541–546 (1984).

[Takagi-93] T. Takagi, E. Miyasaka, "A speech prosody conversion system with a high quality speech analysis-synthesis method," *Proc. Eurospeech93*, pp. 995–998 (1993).

[Takeda-85] S. Takeda, Y. Asakawa, A. Ichikawa, "A study of speech synthesis method utilizing residual informaiton," *Trans. of the committee on speech research, Acoust. Soc, Jpn.*, S84-75, pp. 589–596 (1985).

[Tanaka-93] M. Tanaka, Y. Nomura, Y. Yamashita, R. Mizoguchi, "Automatic generation of speech synthesis rules for accent components based on decision tree (in Japanese)," *Technical report of IEICE*, SP93-59, pp. 15–22 (1993).

[Tohkura-80] Y. Tohkura, Y. Sagisaka, "Synthesis by rules using CV syllable (in Japanese)," *Acoust. Soc. Jpn. meeting*, 3-4-3, pp. 623–624 (1980).

[Watanabe-88] T. Watanabe, "Quality assessment for synthetic speech using word intelligibility score," *The second joint meeting of ASA and ASJ* (1988).

[Watanabe-91] T. Watanabe, "Global assessment method for synthesized speech (in Japanese)," *Trans. IEICE* Vol. J74–A No. 4 pp. 599–609 (1991).

[Yamashita-93] Y. Yamashita, M. Tanaka, Y. Amako, Y. Nomura, Y. Ohta, A. Kitoh, O. Kakusho, R. Mizoguchi, "Tree-based approaches to automatic generation of speech synthesis rules for prosodic parameters," *IEICE Trans. Inf. & Syst.*, Vol. E**76**-A, No. 11 (1993).

Index

Volume 1 (Manufacturing Engineering)
AUTOMOBILE ELECTRONICS
by Shoichi Washino

Volume 2 (Electronics)
MMIC-MONOLITHIC MICROWAVE INTEGRATED CIRCUITS
by Yasuo Mitsui

Volume 3 (Biotechnology)
PRODUCTION OF NUCLEOTIDES AND NUCLEOSIDES BY FERMENTATION
by Sadao Teshiba and Akira Furuya

Volume 4 (Electronics)
BULK CRYSTAL GROWTH TECHNOLOGY
by Shin-ichi Akai, Keiichiro Fujita, Masamichi Yokogawa, Mikio Morioka and Kazuhisa Matsumoto

Volume 5 (Biotechnology)
RECENT PROGRESS IN MICROBIAL PRODUCTION OF AMINO ACIDS
by Hitoshi Enei, Kenzo Yokozeki and Kunihiko Akashi

Volume 6 (Manufacturing Engineering)
STEEL INDUSTRY I: MANUFACTURING SYSTEM
by Tadao Kawaguchi and Kenji Sugiyama

Volume 7 (Manufacturing Engineering)
STEEL INDUSTRY II: CONTROL SYSTEM
by Tadao Kawaguchi and Takatsugu Ueyama

Volume 8 (Electronics)
SEMICONDUCTOR HETEROSTRUCTURE DEVICES
by Masayuki Abe and Naoki Yokoyama

9 789056 990954